Table of Contents

Chapter 1 — page 1
Intro to A&P

Chapter 2 — page 9
Chemical Basis of Life

Chapter 3 — page 14
Care & Use of the Microscope

Chapter 4 — page 16
Cell Anatomy

Chapter 5 — page 24
Tissues

Chapter 6 — page 36
Integumentary System

Chapter 7 — page 41
Skeletal System

Chapter 8 — page 52
Muscle Tissue

Chapter 9 — page 58
Skeletal Muscle Identification

Chapter 10 — page 66
Nervous Tissue

Chapter 11 — page 70
The Brain

Chapter 12 — page 79
Cranial Nerves

Chapter 13 — page 84
Sensory Organs

Intro to Anatomy & Physiology   Lab 1

Introduction

Anatomy and physiology is the complex and contiguous study of the human body and its functions. Learning this material requires good communication skills. You must be able to work with a strong understanding of the language constantly being used in the field. This logical standardized language is both informative and descriptive. Due to the complexity of the vocabulary used in this field, you must first learn to understand its nomenclature. You will then begin to incorporate this new language into the course. You will find that you will use this language throughout both of the required courses of anatomy and physiology, as well as in your chosen healthcare profession.

Descriptive Terminology

Dorsal - toward the back
Ventral - toward the front
Cranial - toward the head
Caudal - toward the tail
Proximal - toward the trunk
Medial - toward the midline
Internal (Deep) - away from the surface

Posterior - toward the back
Anterior - toward the front
Superior - toward the head
Inferior - away from the head
Distal - away from the trunk
Lateral - away from the midline
External (Superficial) - toward the surface

Examples of Terminology:

The eyes are **lateral** to the nose.
The eyes are **medial** to the ears.
The hand is **distal** to the elbow.
The elbow is **proximal** to the hand.
The skin is **superficial** to the muscles.
The bones are **deep** to the skin.
The umbilicus is on the **ventral** portion of the body.
The buttocks are on the **posterior** aspect of the body.

Planes of Reference and Anatomical Positioning

When examining the human body, one must always describe the relationship between one body part and another in as specific a manner as possible, therefore a reference guide describing the body was developed.

Anatomical Position/Anatomical Man

In anatomical position, the body is erect, the feet are parallel to one another and flat on the floor, the eyes are directed forward, and the arms are at the sides of the body with the palms of the hands facing anterior, the thumb pointing laterally and the fingers pointed straight down.

Anatomical Planes

There are three anatomical planes used to describe sections of the body.

> **Median Sagittal Plane**: This plane extends from the front to the back and from the top to the bottom. This plane divides the body into left and right halves.

> **Frontal (Coronal) Plane**: This plane cuts the body into anterior and posterior halves.

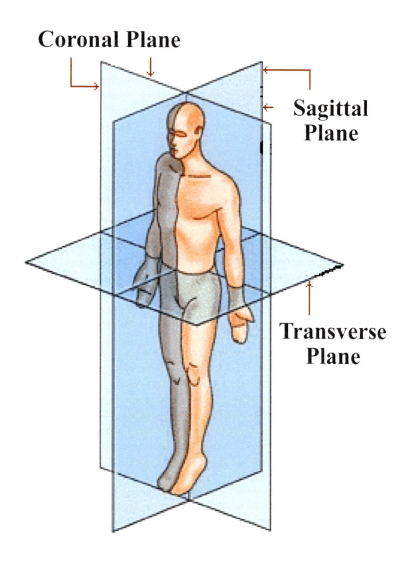

Body Regions

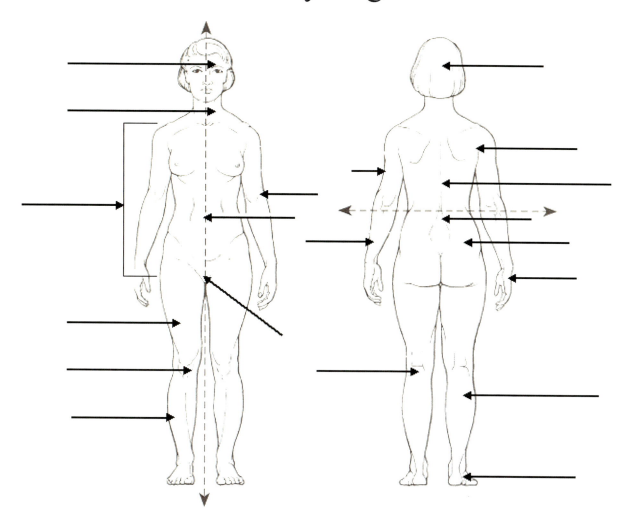

Head: Facial Region: Includes eyes, nose, and mouth
Cranial Region: Covers and supports the brain

Neck: Cervical Region: Supports head and permits its movement

Trunk:
Thorax
Thorax: Referred to as the chest (Thoracic Region)
Mammary Region: Area surrounding breasts
Sternal Region: Area between the Mammary Region
Axillary Fossa: Area of the armpit (Axilla)
Vertebral Region: Area extending down the back
Abdomen
Umbilicus: Center of the abdomen
Abdomen is divided into 4 quadrants and 9 regions
Pelvis
Pubic Area: Symphysis Pubis in genital region
Perineum: Containing external sex organs & anus
Lumbar Region: Back side of the abdomen
Sacral Region: Termination site of vertebral column
Buttock: Posterior aspect also called Gluteal Region

Upper Extremity:

Shoulder: Deltoid Region
Brachium: Upper arm
Cubital Region: Area between arm & forearm
Cubital Fossa: Anterior portion of the elbow
Antebrachium: Area of the forearm
Manus: Carpals: Area of the wrist
Metacarpals: Fleshy portion of hand
Phalanges (Phalanx): Fingers

Lower Extremity:

Hip: Gluteal Region
Thigh: Femoral Region
Knee: Patellar Region (Anterior)
    Popliteal (Posterior)
Leg: Crural Region (Anterior Leg)
    Sural Region (Posterior Leg)
Pes Tarsals: Ankle
    Metatarsals: Fleshy portion of foot
    Phalanges: Toes

The abdominopelvic cavity can be divided into nine regions and four quadrants.

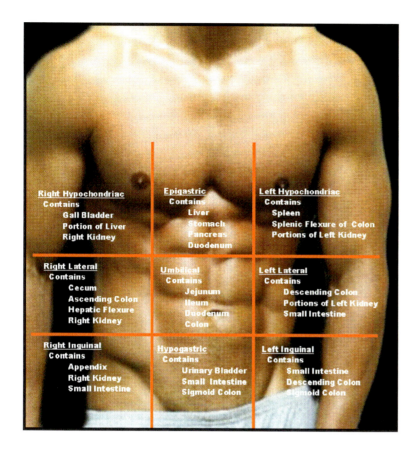

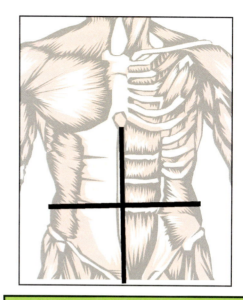

Left Upper Quadrant/Region
Right Upper Quadrant/Region
Left Lower Quadrant/Region
Right Lower Quadrant/Region

Two types of membranes are found in the body. They are composed of thin layers of connective tissue and epithelial tissue that cover, separate, and support the visceral organs. These layers also line the body cavities.

Mucous Membrane: This membrane secretes a thick sticky fluid called mucus. It helps to lubricate and protect the associated organs where it is secreted.

Serous Membrane: This membrane secretes a watery lubricant called serous fluid. This type of membrane forms the pleurae which are membranes associated with the lungs, heart, and the organs of the abdominopelvic region.

    Pleurae: The visceral pleurae adheres to the outer surface of the lung, and the parietal pleurae lines the thoracic walls and the thoracic surface of the diaphragm. The pleural space is located between the two membranes

    Pericardium: The Visceral Pericardium is the serous membrane that covers the heart, and the parietal pericardium forms the outer layer. The pericardial space is located between the two membranes.

    Peritoneum: The visceral peritoneum covers the visceral organs of the abdominopelvic cavity with the parietal peritoneum forming The outer layer. The peritoneal space lies between the two membranes.

Dissection Exercise #1
Examination of the Rat

1) Place your rat on its back and tie the legs apart by tying a string to one leg and wrapping the string around the pan and tying it to the other leg.

2) **Head and Neck Region:**
   a) Beginning inferiorly at the lower mandible of the jaw, make a shallow midline incision continuing down to the superior aspect of the sternum. Make a transverse incision at the neck and the at the superior portion of the sternum. Fold back the skin to allow access to the neck.
   b) Examine the external nares at the anterior nose which are openings into the respiratory system. Air from the nares pass posteriorly and inferiorly into the pharynx. Inferior to the pharynx is the larynx that contains a small opening into the trachea. A small flap of cartilage, the epiglottis covers the opening when the rat swallows to prevent food from entering the trachea.
   c) Examine the thyroid gland, part of the endocrine system, it functions in regulating metabolism. Another endocrine gland is located anterior to the trachea, it is the thymus. This gland functions in the immune system.

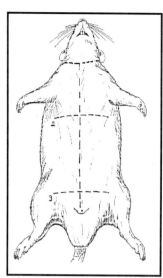

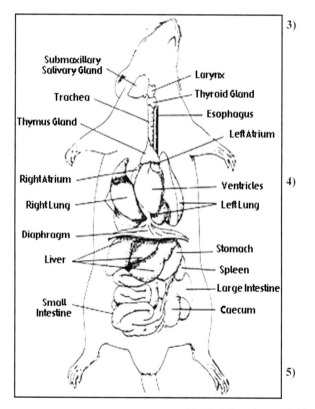

3) **Thoracic Region:**
   Use scissors to continue the incision cutting through the sternum. Make two incisions extending laterally at both the superior and inferior and superior boundaries of the thoracic cavity. Open the rib cage and identify the thoracic structures.
   1) Note the centrally located heart.
   2) Lateral to the heart note the lungs and the centrally located trachea.
   3) Behind the trachea is the esophagus
      Note the inferior boundary of the thoracic cavity, the thin muscular diaphragm.

4) **Abdominal Region:**
   Continue the midline abdominal incision inferiorly to just above the genital region. Make two lateral incisions along the base of the incision to open up the abdominal cavity and note the abdominopelvic structures.
   1) Note the liver, inferior to the diaphragm.
   2) Look posterior to the liver for the gall bladder.
   3) Find the sac-like stomach inferior to the liver.
   4) Find the spleen, inferior to the stomach and to the left.
   5) Small intestine, inferior to the stomach.
   6) Large intestine is continuous with the small intestine
   7) Find the accessory organ the pancreas located inferiorly and slightly lateral to the stomach.
   8) Lift up the small and large intestines and find the kidneys located posteriorly and retroperitoneal.

5) **Pelvic Region:**
   Continue the midline incision inferiorly to the anus.
   1) Find the urinary bladder.
   2) Examine the urethra leading from the bladder.

6) **Cleanup:**
   a) Place all tissue into the trash bin reserved for this dissection. All gloves, paper towels, and practically everything touching the dissection should be trashed in this receptacle.
   b) Dissection tray should be thoroughly washed and dried before returning them to their storage location.
   c) Spray and wipe off your table with disinfectant.

Review Exercises

Identify the following:
- A) Antebrachium
- B) Brachium
- C) Calcaneal Region
- D) Cervical Region
- E) Costal Region
- F) Crural Region
- G) Cubital Region
- H) Frontal Region
- I) Gluteal Region
- J) Lumbar Region
- K) Mammary Region
- L) Manus
- M) Mental Region
- N) Nasal Region
- O) Orbital Region
- P) Patellar Region
- Q) Pelvic Region
- R) Pes
- S) Popliteal Region
- T) Pubic Region
- U) Sacral Region
- V) Sternal Region
- W) Sural Region
- X) Thigh Anterior
- Y) Thigh Posterior
- Z) Thoracic Region

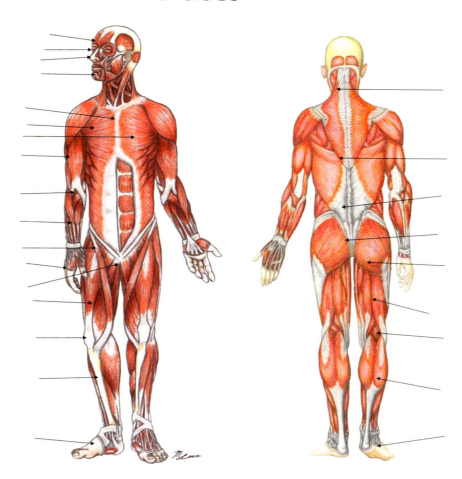

Identify which of the following planes is being used in each picture below:
- A) Midsagittal Plane
- B) Frontal (Coronal) Plane
- C) Transverse Plane

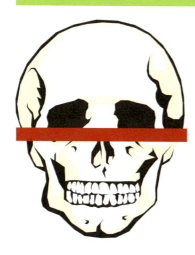 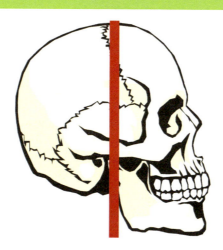 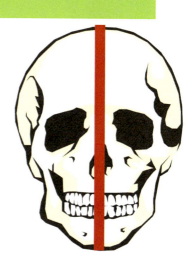

Identify the following statements with the correct anatomical relationship terms.

1) The thigh to the foot _____
2) The hand to the shoulder _____
3) The skin to the bones _____
4) The nose to the mouth _____
5) The nose to the ear _____
6) The thumb to the little finger (fifth digit) _____
7) The lungs to the rib cage _____
8) The sternum to the axilla _____
9) The head to the thorax _____
10) The elbow to the hand _____

Define anatomical position, and explain why standardized positioning is necessary when describing body structures.

Complete the following statements.

1) The _____ cavity contains the brain and the _____ cavity contains the spinal cord.

2) The abdominopelvic cavity is separated by the muscular _____ from the _____ cavity.

3) In the thoracic cavity, the esophagus, thoracic nerves, pericardial cavity, and the heart occupy an area called the _____.

4) The abdominopelvic region is divided into nine regions, the _____ region encases the jejunum, ileum, duodenum, and the colon.

5) A coronal cut of the body divides it into _____ and _____ halves.

6) A sagittal cut of the body divides it into _____ and _____ halves.

Identify the following structures on the picture below.

Abdominal Cavity
Anterior Cavity
Cranial Cavity
Diaphragm
Pelvic Cavity
Posterior Cavity
Spinal Cavity
Thoracic Cavity

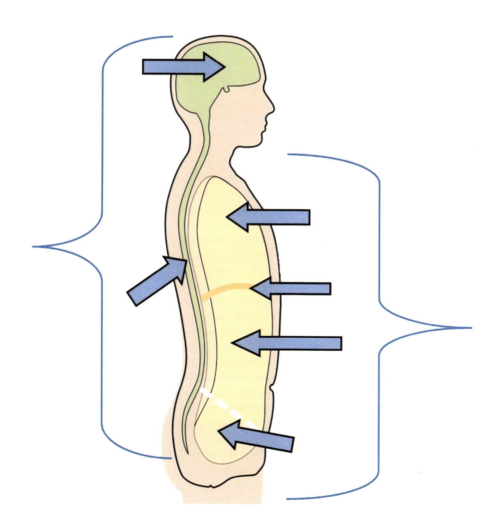

Chemical Basis of Life   Lab 2

Chemistry is not mandatory for the study of anatomy, but it is vital for the understanding of physiology. Physiological functions must have energy to occur and they receive this energy from chemical reactions that occur in the body.

Chemistry Terms

Matter: anything that has mass and takes up space, includes solids, liquids and gasses.

Elements: All matter is composed of elements, and they are its' basic building blocks. There are 118 elements that have been discovered, but depending on your text these numbers range between 111-118 as found on the periodic table of elements. Only 92 occur naturally. Living organisms require about 26 elements. Of these Oxygen composes 65%, Carbon composes 18%, Hydrogen composes 10%, and Nitrogen 3%. These four accounts for approximately 96% of the body.

Atoms: Elements are composed of tiny particles called atoms. These atoms vary in size, weight, and their reactive properties with other atoms.

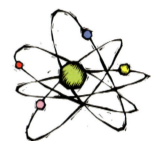

Atomic Structure:

All atoms have a nucleus and one or more electrons. These electrons constantly move around the nucleus. The nucleus itself if made up of one or more particles called protons, and can contain one or more neutrons. **(Subatomic Particles)**

Electrons: Carries a negative charge
Protons: Carries a positive charge
Neutrons: Carries no charge

The nucleus always has protons at its core, and is always positively charged. But, there are usually an equal number of electrons circling the nucleus which allows the atom to be stable. An atom that is not stable will attempt to bind with another atom attempting to reach stability.

Atomic Number: Each atom has its own atomic number which is equal to the number of protons in the atom.
Carbon has 6 protons at its core and its atomic number is 6

Atomic Weight: The total number of protons added with the total number of neutrons is approximately equal to the atomic weight of an atom
Carbon has 6 protons and 6 neutrons and its atomic weight is 12

Isotope: An atom that has a different number of neutrons than protons.

Bonding of Atoms:

When two or more atoms combine, they either gain, lose, or share the electrons between themselves. Electrons of atoms lie in different levels or orbits near the nucleus. These levels or orbits are called shells. The first shell is always closest to the nucleus and holds only two electrons.

First shell - 2
Second Shell - 8   **Octet Rule:** An atom with eight electrons in its outermost shell is
Third Shell - 8   stable.

Ionic or Electrovalent Bonding: When two atoms exchange electrons through electrical charge

Covalent Bonding: When electrons are shared by two atoms

When two atoms of the same element join they create a molecule. When two different elemental atoms combine they form molecules called compounds.

Formulas: Molecular formulas consist of the atoms and their number in the molecule
Glucose: $C_6H_{12}O_6$ 6 Carbon, 12 Hydrogen, and 6 Oxygen atoms

Chemical Reactions

Reactions form or break bonds between atoms, ions, or molecules

Synthesis: When two or more atoms bond to form a more complex structure

$$A + B \longrightarrow AB$$

Synthesis requires energy and is particularly important in the growth of body parts

*In the body we see this in the form of **Dehydration Synthesis** because a molecule of water is removed during the reaction*

Decomposition: When two or more atoms break to form smaller simpler molecules

$$AB \longrightarrow A + B$$

*In the body we see this in the form of **Hydrolysis** because the addition of water causes the molecule to split during the reaction*

Exchange Reaction: When atoms or molecules switch or trade positions with each other

$$AB + CD \rightleftharpoons AC + BD$$

Acids and Bases

Some compounds release ions when they dissolve or react with water molecules

NaCl releases Na+ and Cl- ions when dissolved in water. Due to their ionic charges the solution conducts electricity. Therefore these ions are called electrolytes

Electrolytes that release H+ ions are **Acids**. $HCL \longrightarrow H^+ + Cl^-$

Electrolytes that combine with H+ ions are **Bases**. $NaOH \longrightarrow Na^+ + OH^-$

Electrolyte concentration is of importance and is determined by the **pH Scale** which measures the percent of free H^+ ions in a solution. This scale ranges from 0-14 with 7 being neutral, a proportion of 50% H^+ and 50% OH^- ions. A measurement of less than 7 is considered acidic and a measurement of greater than 7 is considered basic.

The pH Scale

0	1	2	3	4	5	6	7	8	9	10	11	12	13	14
1 M HCl	Stomach Acid	Lemon Juice	Vinegar			Milk	Pure Water / Blood		Milk of Magnesia		Ammonia			1 M NaOH

Acidic — **Basic**

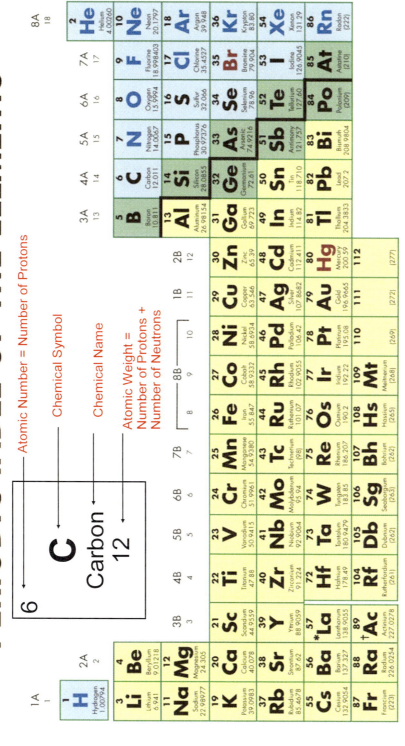

Known 1 _____						
Known 2 _____						
pH	2	4	6	8	10	12
Color	Red	Purple	Violet	Blue	Blue-Green	Greenish Yellow

Procedure

- Chop the cabbage into small pieces until you have about 2 cups. Place it in a large beaker or other glass container and add boiling water to cover. Boil for at least ten minutes allowing for the color to leach out of the cabbage.
- Filter out the plant material to obtain a red-purple-bluish colored liquid. This liquid is at about pH 7. (The exact color you get depends on the pH of the water.)
- Pour about 50 - 100 ml of your red cabbage indicator into a 250 ml beaker.
- Obtain the two known solutions and test each using the above chart as a guide.
- Obtain the six unknowns from your instructor and record your results below.

What PH is considered Acidic

_____

What PH is considered Basic

_____

Unknown	1	2	3	4	5	6
Estimated pH Litmus pH						
Color						
Base/Acid						

1. How many naturally occurring elements are there?

2. What is the basic unit of matter?

3. Identify the four major elements with their approximate percentages.

4. Briefly explain how to find the Atomic Number.

5. Briefly explain how to find the Atomic Mass?

6. Briefly name and describe the three subatomic particles.

7. What is a covalent bond?

8. What is an ionic bond?

9. What is a chemical reaction? Give an example.

10. Identify the differences between a molecule and a compound.

11. Identify the main difference between an organic and inorganic structures.

12. What is the difference between an acid and a base?

13. Give examples of acids and bases.

Care and Use of the Microscope — Lab 3

Introduction:
Anatomy and physiology uses both gross anatomy and microscopic anatomy to study the human body. Gross anatomy deals with specimens one can see with the naked eye, while microscopic anatomy requires the use of a microscope to view the specimens.

The microscopes we use in lab will be of two types:

> **Monocular** - One ocular lens
> **Binocular** - Two ocular lenses

Both are compound microscopes which magnify the image through two lenses (the ocular and the objective lenses). The image you see is of the specimen found on a glass slide that uses light to brightly illuminate the specimen This allows us to classify them as bright field microscopes.

Parts of a Microscope

> **Base:** Solid foundation of the microscope
> **Arm:** Angular portion of the frame that extends upward from the base
> **Stage:** Adjustable platform on which the microscope slide is placed
> **Stage clips:** Metal clips on the stage that holds the slide in place
> **Condenser:** Iris diaphragm located directly under the stage opening
> **Condenser adjustment knob:** Raises or lowers the condenser to alter illumination
> **Lamp:** Light source on the base, below the condenser
> **Coarse Adjustment Knob:** Large knob on the arm used initially to focus specimen
> **Fine Adjustment Knob:** Small knob on the arm to fine focus the Specimen
> **Objective Lenses:** Magnifying lenses of different powers mounted on the nosepiece
> **Ocular Lens:** Eyepiece; lens or lenses through which the specimen is viewed. Ocular lens is always at a magnification of 10X

Microscope Experiment

1) Obtain a specially prepared slide of a letter 'e' from the microscope slide tray.
2) Always begin looking at a slide on the lowest power objective. In this instance it is 4x.
3) Before placing the slide on the stage, lower the stage to its lowest level.
4) Secure the slide on the stage using the stage clips to hold it in place.
5) While looking through the ocular lens, begin raising the stage slowly focusing the slide.
6) Use the fine adjustment knob to bring the specimen into clear focus.
   Note: **Everything viewed through the ocular lens is considered the field of view.**
7) Slowly maneuver the slide side to side and up and down viewing changes in the field.
8) To observe a specimen at high magnification, the following steps must be taken:
   a) Place the specimen in the center of the field of vision.
   b) Obtain sharp focus on the specimen, and then turn the objective ring to the next higher objective until it clicks into place.
   c) Use the fine adjustment knob to bring the structure into sharp focus.
   Note: **You will only need to use the course adjustment knob on the lowest objective, the specimen will continue to be in near focus with each higher objective. You will only need to fine focus the specimen at each higher objective.**
   This feature is called Parfocal.
9) Draw the letter 'e' for each of the objective strengths.

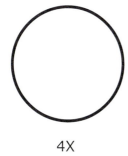 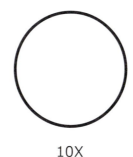 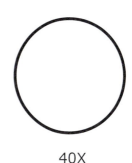

4X   10X   40X

Review Exercise

Identify the following structures of the Microscope

a) Arm
b) Base
c) Condenser
d) Coarse Adjustment knob
e) Fine Adjustment Knob
f) Lamp
g) Objective Lens
h) Ocular Lens
i) Stage
j) Stage Clips

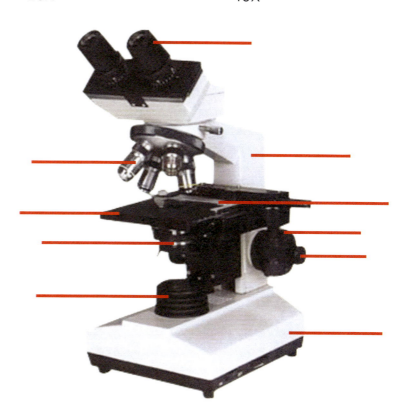

Review Exercises

Explain the magnification process of the microscope.

In your own words define parfocal.

Complete the following statements:

1) The _____ on the stage of a microscope hold the slide in place.

2) The _____ is the upward extension of the base of the microscope by which the microscope should be carried.

3) The light intensity transmitted through the slide is controlled by adjusting the _____ .

4) To view structures through a microscope, you must look through the _____ .

5) When viewing a slide with the 10X objective lens, the microscope is at _____ total magnification.

6) The ocular lens has a magnification of _____ .

Cell Anatomy   Lab 4

The cell is often considered the basic unit of living organisms. Actually, cells are only one of several levels of organization in the human organism:

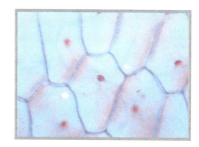

ATOMS
MOLECULES
ORGANELLES
CELLS
TISSUE
ORGANS
ORGAN SYSTEMS
ORGANISM

Thus cells are parts of a larger unit (tissues & organs) and are composed of smaller units (organelles). For now we will concentrate on the structure and function of the cell and its organelles. We will study some of the other levels of organization in later units.

Cell Theory: *The Cell is the Structural and Functional unit of the body and the body is composed of more than one cell.*
Schleiden and Schwann

Parts of the Cell

Cells are microscopic units composed of a bubble of fatty material filled with a water-based mixture of molecules and tiny particles. Parts of a cell are called **organelles** (meaning small organs).

1. The outer boundary of the cell is the **plasma membrane**. It is composed of a double layer (or *bilayer*) of **phospholipid** molecules embedded with other molecules. Each phospholipid molecule has a polar, hydrophyllic ("*water loving*") head made up of phosphate and glycerol; and a nonpolar hydrophobic ("*water fearing*") tail made up of two fatty acid chains. Embedded in the membrane are **integral proteins**, which may have additional protein molecules called **peripheral proteins** attached to them on either side of this membrane. The lipid components account for 98%, the principal component being Phospholipids (75%) and steroid components (23%) while proteins account for only 2%.

2. The double-walled **nucleus** is a large bubble containing the cells genetic code. The code is in the form of **deoxyribonucleic acid (DNA)** strands called chromatin. Portions of chromatin accept stains readily giving the nucleus a very dark appearance.

3. **A nucleolus** (literally, *tiny molecules*) is a small area within the nucleus for the synthesis of ribosomal **ribonucleic acid (RNA).** There may be more than one nucleolus (pleural nucleoli) in a nucleus.

4. The **endoplasmic reticulum (ER)** is a network of membranous tubes and canals winding through the interior of the cell. The **rough ER** is speckled with tiny granules (*ribosomes);* the **smooth ER** is not. The ER transports proteins synthesized within its membrane. The ER also manufactures molecules that make up the cellular membranes.

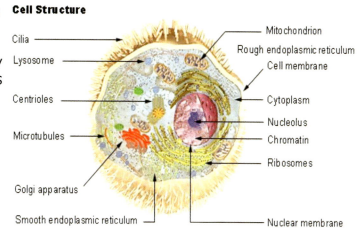

5. The material within the plasma membrane is the **cytoplasm** (literally, *cell stuff*) and includes both the organelles and the liquid, (**cytosol**), surrounding the organelles

6. Tiny bodies that serve as a site for protein synthesis are called **ribosomes** and can be found on the rough ER.

7. The **golgi apparatus** looks like a stack of flattened sacs. It receives products from the ER and packages it into tiny **vesicles** (*bubbles*) for export from the cell.

8. The powerhouse of the cell (due to ATP production) occurs in the **mitochondria**, an oblong capsule with a folded inner membrane called **cristae.**

9. Vesicles that contain digestive enzymes are the **lysosomes** which digest foreign particles and worn out cells.

10. **Microtubules** are very tiny, hollow beams that form part of the supporting cell skeleton, or cytoskeleton. They form parts of other cell organelles (*flagella, cilia, centrioles, and spindle fibers*). Other components of the cytoskeleton are microfilaments and intermediate filaments.

11. The **centrosome** is a dense area of cell fluid near the nucleus. The centrosome contains a pair of centrioles, cylinders formed by parallel microtubules. A network of microtubules called spindle fibers extends from the centrosome during cell division. Spindle fibers distribute DNA equally to the resulting daughter cells.

12. The cell may have any number of other assorted organelles. **Microvilli** are tiny, fingerlike projections of the cell that increase the membrane's surface area for more efficient absorption. **Cilia** are numerous short, hair like organelles that propel material along a cells surface. **Flagella** are single long tails. These organelles found in sperm cells are used for cellular locomotion through the female reproductive tract. **Vesicles** are membranous bubbles that may be formed by the golgi apparatus, or by the pinching inward of the cell membrane to engulf external substances.

Preparing a Microscopic Specimen

We will often use prepared slides of human and other tissues in this course. However, occasionally we may be making our own specimens.

A **wet-mount** slide is a slide on which a wet specimen is placed then covered with a **coverslip. Stains** are used to make a specimen, or some of its parts, more visible. Some require special techniques, but most can simply be added to the specimen and viewed. The usual method for preparing a stained wet-mount slide is described in the following steps.

1. Obtain some skin cells by scraping the inner surface of your cheek with a clean, flat toothpick.

2. Wipe the scrapings on a clean microscope slide and put a small drop of methylene blue stain directly in the smear.

3. After 2-3 minutes rinse the slide carefully ensuring not to remove any of the cellular material.

4. Place one edge of a cover-slip on the slide next to the specimen, then let it drop slowly onto the specimen. This method prevents air bubbles from forming.

5. Absorb any excess fluid around the edges of the cover-slip with a paper wipe.

6. Locate and draw a cell with the low-power objective, shift to high power and repeat.

Sketch your observations below.

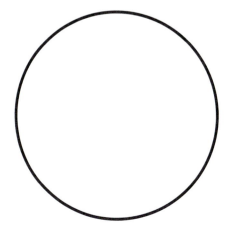

Specimen: *human cheek cells 10X*

Total Magnification: _____

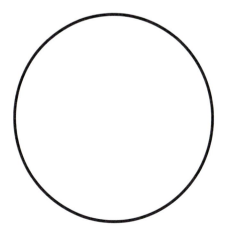

Specimen: *human cheek cells 40X*

Total Magnification: _____

Cell Division

Many of the cells of the human body divide to produce more of the same type cell. The DNA information is carried in the nucleus of the cell which is the genetic coding for that particular type of cell. Before cell division can take place this genetic material must first be duplicated so that the same genetic material is contained in each of the two new cells. Once the genetic material is duplicated the cell undergoes the actual dividing up of all the cellular material. This process of distributing genetic material is the **cell cycle**. The process of cell division actually occurs in a two step process, interphase and mitosis.

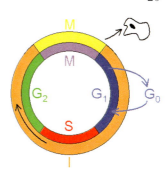

Interphase

This is an inactive period of the cell. It is the growth phase of a new cell that is readying itself for actual cell division and occurs in three stages.
- **G1:** Initial growth phase. In this stage the centrioles begin replicating.
- **S:** The second stage of interphase is the time in which the DNA is replicated
- **G2:** Final stage of interphase; the centrioles finish replicating while the cytoplasm replicates as the cell readies itself for division

Mitosis

The process of mitosis is best described in a step by step process in which we can examine a snapshot of what occurs at specific dividing times. Mitosis occurs in 4 phases: prophase, metaphase, anaphase and telophase.

Prophase: The first phase of mitosis in which the nuclear membrane begins to fragment and disappear. This allows for the chromatin material found within the nucleus to begin condensing and coiling, forming a bar like structure called a **chromosome**. Since DNA replication has occurred during interphase, each chromosome is made up of two identical chromatin threads, now called **Chromatids**. These chromatids are held together at the center by a small spherical body called a **centromere**. The centrioles, which also replicated in interphase, now begin to move to opposite poles of the cell and they act as focal points for the creation of microtubules called **mitotic spindles**. As these spindles grow they push the centrioles toward the opposite poles of the cell. As the nuclear envelope fragments, the spindle fibers interact with the chromosomes and attach to the centromere. These structures are called **Kinetichore microtubules**. The Kinetichore microtubules will be used to pull the chromosomes apart in a later phase of mitosis.

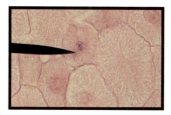

Metaphase: The second phase of mitosis is the point at which the chromosomes are aligned at the exact center or equator of the cell. This arrangement of the chromosomes along a plane midway between the poles is called the **metaphase plate**.

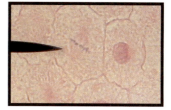

Anaphase: The third phase of mitosis begins at the precise time the chromosomes are split and each chromatid becomes its own chromosome. The Kinetichore fibers pull the newly formed chromosome toward the centrioles which lie at the poles. This phase of mitosis is easily identified because the moving chromosomes look v-shaped.

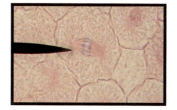

Telophase: The fourth and final phase of mitosis begins when the chromosomes stop moving toward the poles. In essence it is prophase in reverse. The identical sets of chromosomes at the opposite poles of the cell begin to decoil and resume their threadlike chromatin appearance, the nuclear envelope begins to reform around the chromatin mass. The nucleoli reappear within the nuclei and the spindle fibers break down and disappear. At this stage the cell is **binucleate** (*contains two nuclei*) for a brief time and each nucleus is identical to the original mother nucleus. At this time **Cytokinesis**, the dividing up of the cellular contents occurs. A cleavage furrow appears formed by a contractile ring of peripheral microfilaments that squeezes the two new cells apart. Cytokinesis actually begins during late anaphase and continues through and beyond telophase.

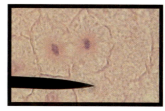

Mitosis Exercise

Retrieve the whitefish mitosis slide from the slide box. On this slide you will see many cells in the different stages of mitosis.

1. With the microscope on a high power find a cell in all the stages of mitosis and draw and label below.

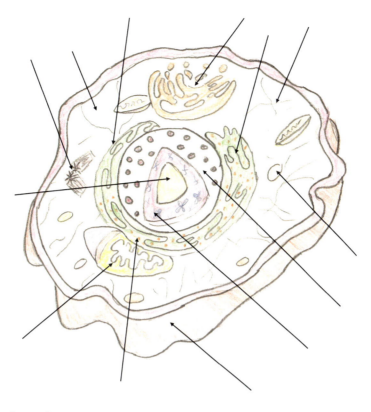

Identify each of structure of the cell using the key below:

A) Centrioles
B) Cytoskeleton
C) Cytosol
D) Golgi Complex
E) Lysosomes
F) Mitochondrion
G) Plasma Membrane
H) Nuclear Membrane
I) Nucleus
J) Nucleolus
K) Ribosomes
L) Rough Endoplasmic Reticulum
M) Smooth Endoplasmic Reticulum

Using the above terms match the below statement.

_____)Composed of a phospholipid bilayer
_____)Organelle of the cell responsible for creating ATP
_____)Actual site of protein synthesis
_____)Membranous sac containing digestive enzymes
_____)Membranous system of tubules with ribosomal attachment
_____)Site of DNA location
_____)Stack of flattened sacs, plays a role in packaging proteins
_____)Provides structural support throughout the cell

Identify the following stages of Cell Division

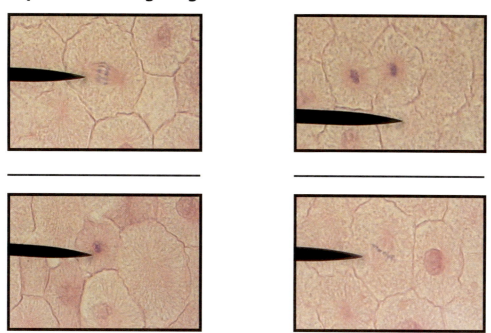

_____   _____

_____   _____

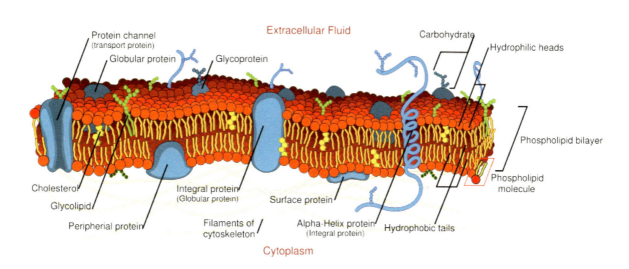

Briefly explain the role of the plasma membrane in the cell.

Tissues   Lab 5

Tissue structures are masses of similar cells and their extracellular matrix. They combine with other tissues to form membranes or organs. The systemic study of these is called *Histology*. When looking at the human body there are four different basic types of tissues in the body:

Epithelial Tissue
Connective Tissue
Muscle Tissue
Nerve Tissue

In each of these different types of tissues there are actual subsets of tissues with differences in structure as well as function. During this lab exercise we will examine several different tissues investigating their structural differences as well as their individual functions.

Epithelial Tissue

Membranous Epithelial tissue consists of tissues that have two basic functions:

Covering Body Structures and Lining Body Spaces
> A key characteristic of this epithelial tissue is that it always has one side exposed, or open to face outward (covering of the body) or inward (lining cavities) of the body. These cells form a continuous sheet, and are held together very tightly with very little matrix (extracellular material). This non free border of the epithelial sheet is attached to underlying connective tissue by a Basement Membrane. The basement membrane is a thin, glue-like layer that holds the epithelium in place while remaining highly permeable to water and other substances. This is a key component of epithelial tissue because epithelia does not have its own blood supply. Water and other important substances must diffuse between the epithelial cells and the underlying tissue through the basement membrane.

Epithelial tissues that form body coverings or body lining is usually classified by structure. The most common classification type is by the number of cell layers and shape of the cell in the outer layer.

Simple Epithelia: The tissue is composed of a single layer of cells.
Stratified Epithelia: This tissue is composed of epithelia that is more than one layer (stratified) of cells.

Simple Epithelial Tissues

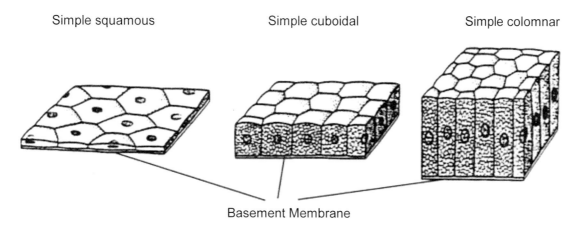

Simple squamous   Simple cuboidal   Simple colomnar

Basement Membrane

Three types of epithelial cells exist: squamous cells are thin and flat, **cuboidal cells** have a square shape in section and a polygonal shape when looking at the surface of a tissue, and **columnar cells** are much taller than they are wide. **Simple epithelial tissues** are only one cell-layer thick and are named on the basis of their cell type: Simple squamous, simple cuboidal and simple columnar.

Stratified Epithelial Tissue

Four types of stratified epithelial cells exist: **Stratified Squamous cells**, **Stratified cuboidal cells**, **Stratified columnar cells** and **Transitional cells**. These are more than one cell-layer thick and are named on the basis of their cell type. Stratified epithelial tissue is very active at the basement layer where the cells are dividing constantly. As the cells continue to divide they push the new cells up toward the outer layer of the tissue. This tissue can be either ciliated (*a border of small hair-like structures on the outer surface of the tissue*) or non-ciliated. This ciliated border helps to move substances across the surface of the tissue. Stratified Squamous tissue can also be keratinzed (*filled with a waterproof material called Keratin which dehydrates and flattens the cells to aid in protection of the body*) or non-keratinized (*cells that maintain their ability to absorb and secrete substances across the surface of the tissue*).

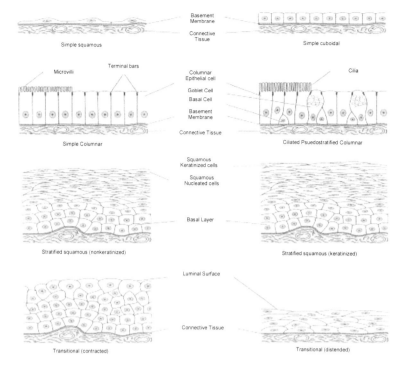

Pseudostratified epithelial Tissue

The name Pseudostratified actually means "false stratified". This tissue appears to be many layers thick, but in reality is composed of a single cell layer. This illusion is created by the way the cells seem to be pushed together. This moves the nucleus of the cell away from the basement membrane and into the upper portion of the cell producing this stratified appearance

Examples of Major Membranous Epithelial Tissues

Tissue	Location	Function
Simple Squamous	Alveoli of Lungs	Absorption by diffusion of respiratory gases between alveolar air and blood
	Lining of blood and lymphatic vessels (endothelium)	Absorption by diffusion, filtration, and osmosis
	Surface layer of pleura, pericardium, and peritoneum (mesothelium)	Absorption by diffusion, osmosis; Secretion
Stratified Squamous Non-keratinized	Surface of mucous membrane lining mouth, esophagus, and vagina	Protection
Keratinized	Surface of skin (epidermis)	Protection
Simple Cuboidal	In glands and ducts, also found in tubules of other organs, as in the kidney	Secretion; Absorption
Pseudostratified	Surface of mucous membrane lining the trachea, large bronchi, nasal mucosa, and parts of the male reproductive tract	Protection
Simple columnar	Surface layer of mucous lining of stomach, intestines, and parts of respiratory tract	Protecion; Secretion; Absoprtion; Movement Mucus (ciliated columnar)
Transitional	Surface of mucous membrane lining urinary bladder and ureters	Permits distension

Creating Glands

Glandular epithelium is a type of tissue that forms glands. This epithelium forms the functional portions of **Exocrine glands** (glands that secrete substances into ducts emptying onto epithelial surfaces), and **Endocrine glands** (glands that secrete substances that diffuse into the bloodstream).

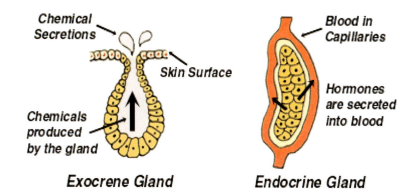

Exocrine Glands:

These glands can be classified by structure. **Unicellular glands** are single cell glands. One example of this is the Goblet Cell. **Multicellular glands** are composed of both secretory cells as well as cells that form the walls of the ducts. They can structurally be either simple (non-branching glands) or compound (branching glands). These glands produce their products and secrete them through ducts. They are classified by the method in which they release their products. **Merocrine Glands** diffuse their products through the Cell membrane and include the **Salivary Glands, Pancreatic Glands,** and the **Sweat Glands. Aprocrine Glands** store their product on the surface of the secretory cell and then a portion of the cell along with its secretion is pinched off and discharged into the body. Mammary glands are of this type. **Holocrine Glands** produce their product and then discharge the entire cell product and all into the blood stream, **Sebaceous Glands** are of this type.

Endocrine Glands:

These glands produce hormones that are released into the bloodstream traveling through the circulatory system to reach their target cells.

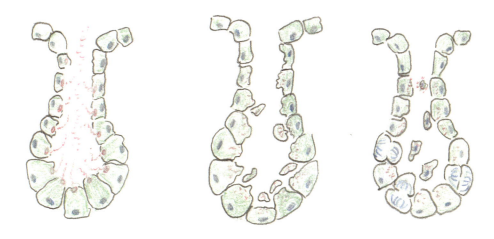

Microscope Exercise

Using the correct microscope procedure retrieve the prepared tissue slide from the slide tray and draw each of the following:

1) **Simple Squamous Epithelium**   Simple Squamous is a single cell layer of flattened mosaic structured cells. Because it is so thin it is best adapted for diffusion or filtration.

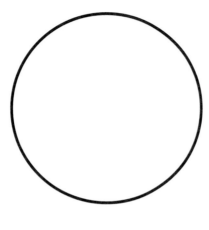

400X

2) **Stratified Squamous Epithelium**   Stratified Squamous is composed of multiple layers of cells: Columnar Cells along the basement membrane, topped by cuboidal cells, then by squamous cells. Cells divide along the basement membrane and are pushed upward where they flatten out.

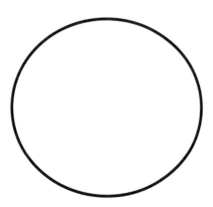

400X

3) **Simple Cuboidal Epithelium**
Simple Cuboidal is composed of a single layer of almost cube like cells. This tissue is found in secreting organs such as glands. It is also found forming the Kidney Tubules where it is specialized for water reabsorption.

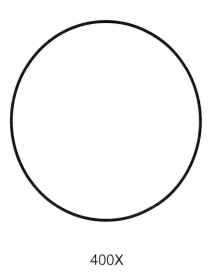

400X

4) **Simple Columnar Epithelium**
Simple columnar is specialized for absorption and secretion and is found in many parts of the body. It lines portions of the reproductive tract, digestive tract, excretory ducts and respiratory tract. A special cell found in simple columnar tissue is the Goblet Cell, a cell that secretes mucus which aids in lubrication and protection of the epithelial lining.

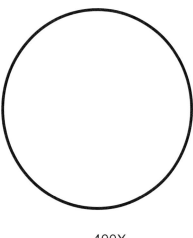

400X

5) **Ciliated Pseudostratified Epithelium**
Ciliated Pseudostratified tissue is a single layer of columnar cells that all attach to the basement membrane. The Nuclei of each cell can lie at differing heights in the cells creating a stratified appearance. Like simple columnar it can be both ciliated and non–ciliated. It can be found in the same areas as simple columnar; the superior esophagus, the upper respiratory tract, and parts of the male reproductive tract.

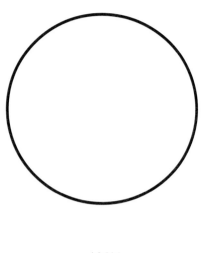

400X

6) **Transitional Epithelium**
Transitional epithelium tissue is specially adapted for stretching and is found in areas of the body that permit distension. This tissue is primarily found in the Bladder.

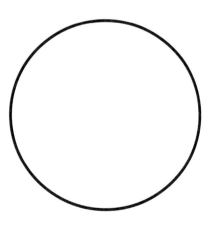

400X

Connective Tissue

Connective tissue is specialized in that it is the connecting portion among various different tissues. Bone, Cartilage and Fibrous connective tissues help to cement tissues together or to support their overall structure. Blood tissue actually helps to connect differing tissues together through the transportation of substances between them.

Classification of Connective Tissues
A commonality of all connective tissues is that they each have an extracellular matrix, or extracellular fluid which separate the individual cells. All Connective Tissue cells are separated by one of three basic types of matrix.

Protein Matrix
This is an extracellular matrix predominately composed of protein fibers. **Collagen**, a common protein forms bundles of tough, flexible fibers. These fibers are sometimes called white fibers due to their whitish appearance. **Elastin** is thick protein fiber that has the ability to stretch over 1 and half times its original length and return to its original shape. These fibers are sometimes called yellow fibers due to their yellowish appearance.

Fibrous Tissues:
This tissue is considered either dense or loose depending on the density of protein found in the matrix. These dense fibrous tissues can be either **Dense Regular** Fibrous tissue (made of regular parallel bundles of fibers) or **Dense Irregular** Fibrous tissue (made of irregularly bundled fibers). **Loose** fibrous tissue is also called **Ordinary Areolar Tissue**.

Adipose Tissue:
Adipose tissue is also called fat tissue due to its primary function of storing fat. This tissue is a modified form of areolar tissue with fat storing cells called adipocytes. These cells store lipids for later use in the body.

Ground Substance or Extracellular Matrix
This extracellular matrix is composed of protein fibers, but it also has an abundance of other non-fibrous proteins and other substances.

Cartilage:
Cartilage is created from a combination of different fibers and ground substance which allows it to have an elastic rubbery consistency. **Hyaline Cartilage** is the most common type of cartilage in the body and has a moderate amount of collagen fibers in its matrix. **Fibrocartilage** is a dense cartilage and is more durable in the body. It has a large amount of collagen fibers in its matrix. **Elastic Cartilage** is very distinct in that it is composed of a large number of elastin fibers allowing for its elastic quality.

Bone:
Bone is composed of a dense matrix of collagen fibers with mineral crystals embedded in its ground matrix. This gives it a solid, hard consistency. There are two types of bone tissue. **Compact bone** forms large very dense layers of boney matrix, and **Cancellous** (*spongey*) **bone**, also called **Trabeculated** bone forms a thin narrow lattice of bone for the storage of Red Bone Marrow

Extracellular Fluid Matrix
This extracellular matrix is a water-based solution with a fluid consistency.

Blood:
Blood is the major type of fluid connective tissue. Suspended within the fluid are three different types of formed elements, **Red Blood Cells** (Erythrocytes) 4.2-5.8 million per cubic milliliter, **White Blood Cells** (Leukocytes) 5000-9000 per cubic milliliter, and **Platelets** (Thrombocytes) 250,000-400,000 per cubic milliliter. Plasma accounts for 55% of blood while the Formed Elements accounts for 45%

Hemopoetic Tissue:
Hemopoiesis is the production of blood cells. This tissue, called *myeloid tissue* forms the Red Bone Marrow, a primary site for red blood cell production.

Microscope Exercise
Using the correct Microscope procedure Retrieve the prepared tissue slide from the slide tray and draw each of the following:

1) **Cartilage**
Cartilage consists of cells called chondrocytes. It is a semisolid matrix that allows for marked elastic properties of the tissues. It forms a precursor to one type of bone and persists at the articular surfaces (movable ends) of bones. Chondrocytes occupy a cavity called a Lacunae within the matrix. Mature adult cartilage is avascular.

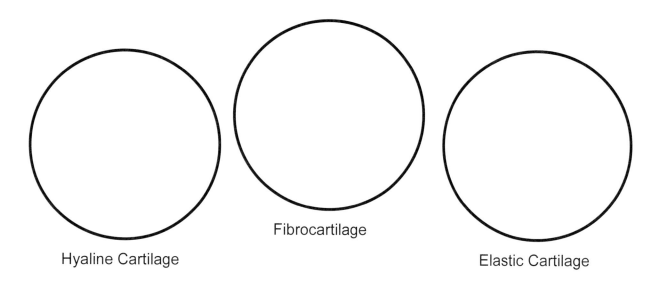

Hyaline Cartilage   Fibrocartilage   Elastic Cartilage

2) **Bone**
Bone is the most rigid of all connective tissues. Unlike cartilage, it has a rich vascular supply. The hardness of bone is largely due to the calcium phosphate (calcium Hydroxyapatite) deposited in its matrix. Numerous collagenous fibers are also embedded in the matrix to give the bone some flexibility. Bone cells (Osteocytes) are arranged in concentric layers (Lammellae) around a central canal. Each osteocyte occupies a cavity called a lacunae. Radiating from each lacunae are numerous miniature canals called Canaliculi.

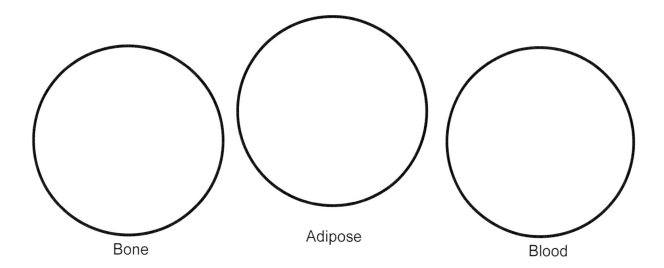

Bone   Adipose   Blood

Microscope Exercise Continued
Using the correct Microscope procedure Retrieve the prepared tissue slide from the slide tray and draw each of the following:

3) Blood
Blood plays a vital role in Homeostasis. This fluid matrix contains the three formed elements, Red Blood Cells, White Blood Cells, and Platelets.

4) Muscle Tissue
Muscle is responsible for the movement of materials through the body. It also allows for the physical movement of the body. Muscle is divided into three distinct types, Smooth Muscle, Skeletal Muscle, and Cardiac Muscle. Smooth Muscle has fibers that are long and spindle shaped containing a single nucleus. This muscle type lacks the striated appearance of other muscle types and is considered involuntary. Skeletal Muscle attaches to the skeleton and is responsible for body movement. Fibers of this muscle tissue are grouped together into parallel fasciculi and have multiple nuclei within the cells. This muscle tissue is considered voluntary. Cardiac Muscle is found only in the heart and looks very similar to skeletal muscle with two exceptions. The muscle fibers are branching fibers with a single central nucleus and contain intercalated discs.

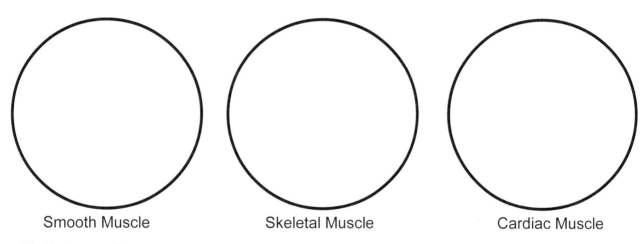

Smooth Muscle   Skeletal Muscle   Cardiac Muscle

5) Nervous Tissue
Nervous tissue is composed of neurons, which respond to stimuli and conduct impulses to and from the brain. The three main parts of the neuron are the dendrite, the cell body (*Perikaryon*), and the axon. Dendrites are branched processes that receive stimuli and conduct impulses to the cell body which contains the nucleus and other specialized organelles. The cell body then transmits the impulse away from the neuron via the axon.

Draw and label a neuron in the space below.

Review Exercises

Name of Tissue	Location of Tissue	Function of Tissue
Fibrous Loose, Ordinary Areolar	1) _____	Connection
Adipose (Fat)	Under Skin Padding at various locations	2) _____
Dense Fibrous Regular	3) _____	Flexible strong connection
4) _____	Deep Fascia, Dermis, Scars	Connection, support
Bone 5) _____	Skeleton	Support, Protection, Calcium Reservoir
Cancellous Bone	Skeleton	6) _____
Cartilage Hyaline	7) _____	Firm Flexible Support
8) _____	Disks between vertebrae, Symphysis Pubis	Firm Flexible Support
Elastic Cartilage	9) _____	Firm Flexible Support
Muscle Smooth Muscle	10) _____	Involuntary Movement of Substances in the body.
11) _____	Attached to Skeleton	Voluntary Movement of the Body
Cardiac Muscle	Heart	12) _____

Review Exerecise

Use the key to identify the major tissue types described below.

Key: A) Connective Tissue   B) Epithelium   C) Muscle   D) Nervous Tissue

1) _____ Lines body cavities and covers the body's external surface

2) _____ Pumps Blood, Enables Movement, Propels Food

3) _____ Transmits Electrical impulses

4) _____ Used in absorption, secretion, and filtration

5) _____ Major function is to contract

6) _____ Most involved in regulating and controlling body functions

7) _____ Can contain Keratin

8) _____ Forms exocrine and endocrine glands

9) _____ Forms the nerves and the brain

10) _____ Forms the intervertebral disk

Briefly explain the differences between simple and stratified epithelium.

Briefly explain the differences between Endocrine and Exocrine Glands.

Briefly explain the special characteristic of Transitional Epithelium.

Review Exercises

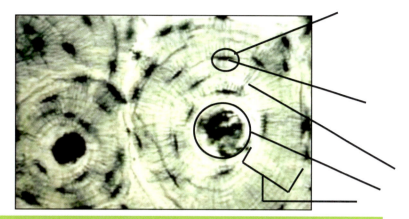

Using the Key below Identify the above Structures

A) Central Canal
B) Osteocyte
C) Canaliculi
D) Circumferential Lamellae
E) Lacunae

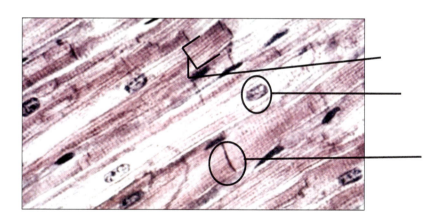

Using the Key below Identify the above Structures

A) Nucleus
B) Intercallated Disk
C) Muscle Striations

Integumentary System   Lab 6

The integument includes the skin and its accessory organs; hair, nails and glands.

The skin is the largest organ in the body and consists of several types of tissues that function together to protect it from external forces helping to maintain homeostasis. The average adult is covered in approximately 7,600 square centimeters or 3,000 square inches of skin. It constitutes approximately 7% of the total weight of the body. Our skin has a variable thickness with an average of approximately 1.5 mm, and has two distinct layers. The outer **epidermis** is composed of 4-5 stratified squamous cell layers, and the inner dermis actually connects the skin to the body.

Major Functions of the skin include:
Physical Protection
Hydroregulation
Thermoregulation
Cutaneous Absorption
Synthesis
Sensory Reception
Communication

Layers of the Epidermis

The **Stratum Basale,** the first layer of the skin lies in direct contact with the dermis. It is highly mitoticly active and has four types of cells, keratinocytes, melanocytes, tactile cells, and langerhan cells. Keratinocytes are cells that create keratin, an enzyme that helps to strengthen and toughen the skin. The protects against abrasion, cuts and excessive pressure. Melanocytes are cells that produce melanin, a pigment of the skin which helps to protect us from ultraviolet light. Tactile cells (*merkels cells*) produce the tactile sensations of the body and Langerhan Cells (Dendritic Cells) are macrophages that aid with protection from bacterial invasion.

The **Stratum Spinosum**, the second layer actually contains several stratified layers. This layer appears to have tiny spine like fibers extending from the Keratinocytes. The Stratum Basale and the Stratum Spinosum are combined to form the Stratum Germinativum.

The **Stratum Granulosum** consists of three to four layers of flattened cells. These cells are filled with keratohyalin which is the precursor to keratin.

The **Stratum Lucidum** is a clear layer of cells only found in thick areas of the skin. This layer is found in the thickened skin of the soles of the feet and the palms of the hand. This layer does not contain any organelles, nuclei, or cell membranes.

The **Stratum Corneum** is composed of 25-30 layer of flattened scale like cells. This is the outer most portion of the skin. These cells are dead cells and are sloughed off during daily activities. Thousands of these cells are lost and replaced each day. This layer is **cornified**, which means that the cells have become flat and dry an important adaptation that provides protection to the body.

The dermis is deeper than the epidermis and is much thicker. Elastic and collagenous fibers are arranged in the dermis to provides lines of tension, which help to give the skin its overall tone. As we age we lose the elastic quality of our skin which causes wrinkling and enables tissue damage from incidental accidents to increase with age.

Layers of the Dermis

The **Stratum Papillare** is the layer directly beneath the epidermis. Numerous papillae, or projections extend from the upper portion of the dermis into the epidermis. Papillae from the base form the friction ridges on the fingers and toes giving us our fingerprints.

The **Stratum Reticulare** is deeper and thicker than the dermis. Fibers of this layer are more regular and dense and form a tough meshwork. This meshwork can be broken and repaired, scars and stretch marks are an indication of this tearing. Stretch marks and scars are often called **Striae**, Older scar damage is often called **Striae Albicans** due to their silvery white appearance.

The hypodermis or subcutaneous tissue is not actually part of the skin, but it aids in binding the skin to the underling organs. It is composed primarily of loose connective tissue and adipose cells, it is interlaced with blood vessels. The amount of adipose tissue varies with the body location and the sex of the individual. Females carry about an 8% thicker hypodermis than males.

Finger Print Exercise

Collect an ink pad from your instructor: Roll your index finger in ink then over the boxed area. Repeat the exercise with your partner's finger. Examine both fingerprints for distinct differences.

Epidermal Derivatives

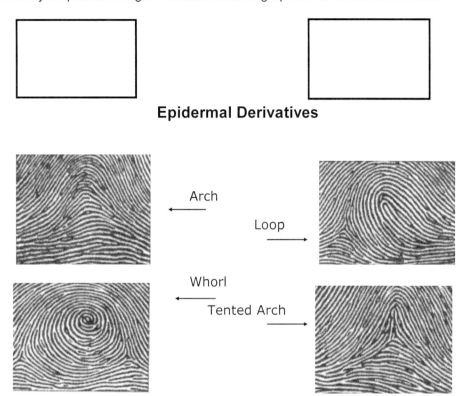

Arch
Loop
Whorl
Tented Arch

Hair

Hair is a distinguishing feature of mammals. humans have a small amount of hair compared to other mammals. Usually dense hair is found on the scalp, face, pubis, and axillae. Men and women have about the same amount of hair on their bodies but due to male hormones, males appear hairier.
The primary function of hair is for protection. This is limited to the areas hair is found on the body and is usually a covering to protect the skin form Ultraviolet Light. Each hair consists of a diagonally positioned shaft, root, and bulb, and develops from a hair bulb located in the Stratum Basale. The bulb receives its nutrients from the dermal blood vessels coursing through this layer of skin. The life span of each hair is different depending no where the hair is located on the body. An eyelash will fall out after 3-4 months of growth, which is in stark contrast to the 3-4 years that a hair from the scalp can continue to grow. The color of hair is determined by the amount of melanin that is produced by the melanocyte. The more melanin produced, the darker the color of the hair. As we age the melanocytes produce less and less melanin creating gray or white hair in old age. The red hues to hair is dependent on an iron based pigment (*Trichosiderin*) found in the body. When looking at the hair in a cross section one can determine the if the hair is straight, wavy, or curly. Straight hair is round, wavy hair is oval, and curly hair is flat.

Distinct Hair Types:

Lanugal Hair is a fine silky fetal hair that appears during the last trimester of development. It can be found on premature infants. People afflicted with Anorexia Nervosa have reappearance of this fetal hair.
Angora Hair grows continuously and is found on the scalp and the face of men. This type of hair continues to grow 3-4 years before falling out. **Terminal (definitive) Hair** grows to a certain length and then stops growing. This type of hair usually lasts 3-4 months and then falls out. Eyelashes, eyebrows, pubic hair, and axillary hair are of this type.

Two Point Discrimination Exercise

Using calipers or an esthesiometer and a metric ruler, test the ability of the subject to differentiate two distinct sensations when the skin is touched simultaneously at two points. Begin the exercise on the face. Start with the caliper arms together and begin the test. After each test gradually increase the distance between the two points until the subject reports that he/she can feel two distinct points. This measurement, the smallest distance at which the two points can be distinguished is the two-point threshold. Repeat with each of the areas in the table below.

Body Area Tested	Two-Point Threshold (mm)
Face	
Back of Hand	
Palm of Hand	
Back of Neck	
Fingertip	

Hair Exercise:

Remove a hair from one of your lab partners. Examine the hair noting its color, its texture, and its shape. Place the hair under the microscope and draw the image in the space below.

Color: _____

Shape: _____

Nails

Nails are formed from the **stratum corneum** of the epidermis. They are extremely dense and hard due to the dense keratin fibrils running parallel to the cells. Each nail consists of a body, a free border, and a hidden border, The nail rests on an area of the finger called the nail bed. The sides of the nail body are protected by a nail fold called the **Paronychium**, and the furrow between the sides of the body is the nail groove. Part of the nail extends over the finger, which is the portion of the nail that is trimmed. It is called the **Hyponychium** or the quick. The proximal portion of the nail is covered by the hidden border of the nail and is called the **Eponychium** or the cuticle of the nail. The growth of the nail occurs in the nail matrix which extends toward the distal aspect of the nail. This growth area leaves behind a small part of the nail that looks like a small white half moon called the **Lunula**. Nails on the fingers grow at an average rate of 0.1mm a day (*1 cm every 100 days*), while nails on the toes grow at a slower rate (1/2 cm/100days) Fingernails require 4 to 6 months to re-grow completely while Toenails require 12 to 18 months. Actual growth is dependent upon age, season, exercise level, and hereditary factors.

Nail Exercise:

Identify the following aspects of the nail:

Hyponychium
Nail Body
Eponychium
Lunula
Nail Root
Paronychium

Glands

The glands of the skin course through the epidermis, but all of the glands of the skin are actually located in the dermis and release their products through ducts which lead to the outer portion of the skin. There are three basic types of glands in the skin. **Sebaceous glands** produce oil or **Sebum**. These glands are found near hair follicles and helps to lubricate the hair, preventing it from becoming brittle which helps to reduce breakage. Acne is caused by the blockage of a Sebaceous Gland. **Sudoriferous glands** produce sweat. Sweat is composed of water, salts, urea, and uric acid. These glands are found mostly on the palms of the hands, soles of the feet, axillary region, pubic region, and the forehead. **Ceruminous glands** produce Cerumen or ear wax. This is water and insect resistant and helps to keep the Tympanic Membrane pliable.

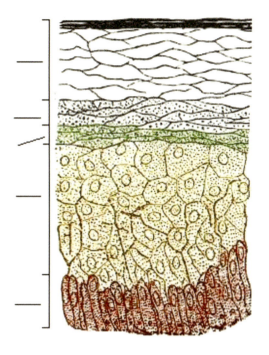

**Stratum Basale
Stratum Corneum
Stratum Granulosum
Stratum Lucidum
Stratum Spinosum**

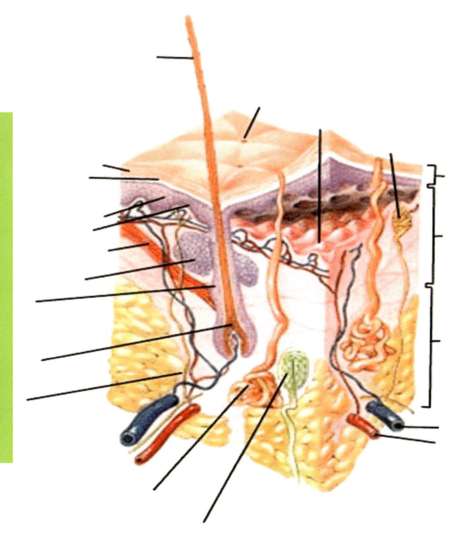

**Arrector Pili Muscle
Artery
Dermal Papillae
Dermis
Epidermis
Hair Follicle
Hair Papillae
Hair Shaft
Hypodermis
Meissner's Corpuscle
Nerve Fiber
Pacinian Corpuscle
Sebaceous Gland
Skin Pore
Stratum Basale
Stratum Corneum
Stratum Granulosum
Stratum Spinosum
Sudoriferous/Sweat Gland
Vein**

Match each of the following terms with their corresponding location or function

1) Stratum Lucidum
2) Ceruminous Gland
3) Melanocyte
4) Lanugo Hair
5) Paronychium
6) Stratum Germinativum
7) Papillary Layer of Dermis
8) Linea Albicans (Striae)
9) Keratinocyte
10) Langerhan /Dendritic Cell
11) Trichosideran
12) Angora Hair
13) Hyponychium
14) Sudoriferous Gland
15) Exocrine Gland
16) Stratum Basale
17) Stratum Corneum
18) Merkel Cells
19) Hypodermis
20) Terminal (Definitive) Hair
21) Eponychium
22) Sebaceous Gland
23) Lunula
24) Epidermis

_____ Macrophages that eliminate bacteria
_____ Type of hair that grows for 3-4 months and then stops growing
_____ Composed of 25-30 layers of flattened cells (outer layer of epidermis)
_____ Type of gland that secretes its product through ducts
_____ This cell gives the skin its color
_____ The subcutaneous structure that attaches skin to the underlying fascia
_____ Nail fold that protects the sides of the nail
_____ Outermost layer of the skin
_____ Type of hair found only on premature infants
_____ Gland that produces ear wax
_____ Created as a result of tissue damage forming scars or stretch marks
_____ The hidden border of the nail, also called the cuticle
_____ This hair grows for 3-4 years, found on the scalp
_____ Layer of the epidermis that undergoes mitosis
_____ Cell that produces Keratin
_____ Composed of both the Stratum Basale and the Stratum Spinosum
_____ Gland when blocked creates Acne
_____ Tactile Cells of the skin
_____ Small half moon structure at the proximal end of the nail
_____ Layer of skin responsible for the friction ridges on the fingers and toes
_____ Gland that produces a liquid composed of water, salts, urea, and uric acid
_____ Free border of the nail, also known as the quick
_____ Responsible for the red hues in skin and hair
_____ Layer of skin found only in thick skin

Complete the following:

1) The _____ is the largest organ of the body.
2) The skin is composed of two layers an _____ composed of stratified squamous cells arranged in 4-5 layers, and a deeper _____ which connects the skin to the underlying organs.
3) The _____ produces melanin which provides protection from the sun's _____ light.
4) The three shapes of hair can be determined by looking at a cross section of the hair fiber to see if it is round, oval, or flat. Round hair fibers are found on _____ hair, and the oval fibers are found on _____ hair, while flat hair fibers are found on _____ hair.
5) Fingernails grow at a rate of _____, while toenails grow at a rate of _____ every 100 days.
6) _____ hair is found on newborns, but can also be seen on people who suffer from _____ disease.
7) _____ is produced in the ear and is _____ and _____ resistant.
8) Exocrine glands are found in the _____ and secrete their products through _____.
10) _____, _____, and _____ are all functions of the skin.
11) The skin's accessory organs include the _____, the _____ and the _____.

The Skeletal System Lab 7

Skeletal system is divided into two components, an **Axial** and **Appendicular Skeleton**. The axial and appendicular skeleton complete a full adult skeleton of approximately 206 bones. These bones are arranged to form a strong yet flexible framework for the body. At birth we have approximately 270 bones in the body, but through growth and ossification processes of these boney structures their number decrease to the approximate number in the mature adult skeleton. Each individual bone plays an important role in the function and structure of the body.

The axial skeleton is composed of those bones that form a protective barrier for the body. Each of these bones are centrally located and protect specific visceral organs. The **skull** protects the brain as well as special sensory organs. The **vertebral column** continues this protective function of the nervous system by surrounding and protecting the spinal cord. The **rib cage** surrounds and protects all of the internal visceral organs. There are two exceptions to this rule. The **hyoid** bone is found in the cervical region and is the only bone that does not articulate with another bone in the body, and the **auditory ossicles**, found in the middle ear cavity of the Temporal bone. Due to their proximity to the skull they have been included here.

The appendicular skeleton is composed of those bones that give rise to extrinsic movement of the body as well as locomotion. It consists of the **pectoral girdle** which connects the upper extremity to the body and the **pelvic girdle** that connects the lower extremities.

As stated above, one of the most important functions of bone tissue is protection, but there are other important functions as well. The skeleton is also used for **structural support** of the body. It forms a rigid framework to which the softer tissues and organs are attached. It is an interesting fact that our skeleton supports a mass of muscles and organs that can weigh as much as five times that of the bones themselves. **Movement**, although not specifically a function of bones, does play an important role in connecting skeletal muscles. They provide anchoring attachment sites for most skeletal muscles that contract for movement. As the muscle contracts the bone serves as a lever creating movement. **Hemopoiesis**, another important function, is the process of blood cell formation. This takes place in the **medullary cavity** where the red bone marrow is housed. Estimations of approximately 100 million red blood cells are produced every minute by this tissue to maintain a constant supply for the body. Another important function of Bone tissue is **mineral storage**. The inorganic matrix of bone is composed primarily of the minerals Calcium and Phosphorus (Calcium Hydroxyapatite). These two minerals constitute 2/3rds of the weight of bone and help to maintain its strong tensile strength.

Skeletal Vocabulary

Articulating Surfaces	
Condyle	Large Rounded Articulating Knob
Facet	Flattened Shallow Surface
Head	Prominent Rounded end of Bone
Depressions or Openings	
Alveolus	Deep Pit or Socket
Fissure	Narrow Slit-like Opening
Foramen	Rounded opening through the Bone
Fossa	Flattened or Shallow Space
Sulcus	Groove that houses a vessel or Nerve
Nonarticulating Prominences	
Crest	Narrow Ridge-Like Projection
Epicondyle	Projections adjacent to a Condyle
Process	Any marked Bony Prominence
Spine	A Sharp Slender Process
Trochanter	Massive Process on Femur
Tubercle	Small Rounded Process
Tuberosity	Large Roughened Process

Shapes of Bones

Long Bones: These bones are longer than they are wide. They function as levers and are chiefly located in the upper and lower extremities.

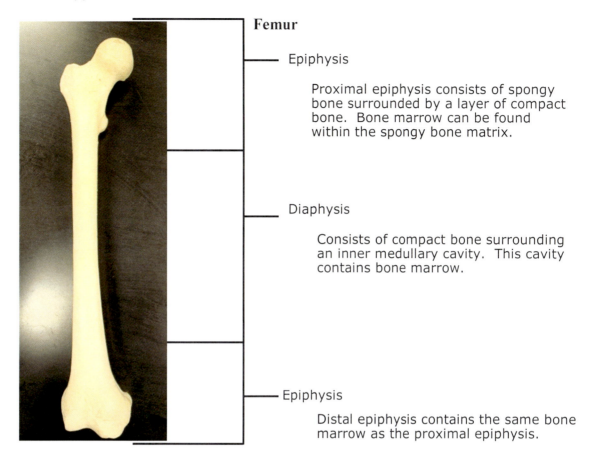

Femur

Epiphysis

Proximal epiphysis consists of spongy bone surrounded by a layer of compact bone. Bone marrow can be found within the spongy bone matrix.

Diaphysis

Consists of compact bone surrounding an inner medullary cavity. This cavity contains bone marrow.

Epiphysis

Distal epiphysis contains the same bone marrow as the proximal epiphysis.

Short Bones: These bones are somewhat cube shaped. They help in the transmission of forces placed on the body and are found in the wrists and ankles.

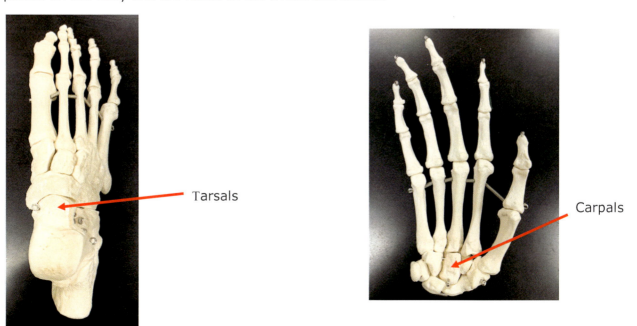

Tarsals

Carpals

Flat Bones: These bones have a broad surface for muscle attachment. They are found in the skull, the ribs and the shoulder girdle. They are chiefly used for protection.

Irregular Bones: These bones are varied in shape and have many surfaces. They are found in the vertebrae and some of the smaller bones in the skull. They are used as muscle attachment sites to increase structural support.

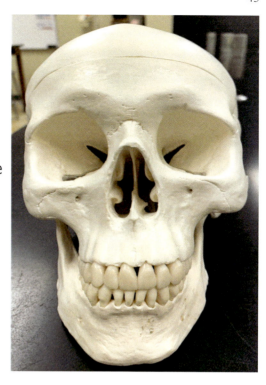

Bone Cells

There are five types of principle bone cells that create and maintain bone tissue. **Osteogenic** cells are located between the **endosteum** and the **periosteum**. They respond to trauma in bone tissue and proliferate into four distinct bone forming cells. **Osteocytes** are mature bone cells. They help in maintaining healthy bone tissue by secreting enzymes, influencing bone mineral content. **Osteoblasts** are bone forming cells that are concentrated under the periosteum and areas bordering the medullary cavity. These cells actually build bone tissue by removing calcium from the body and placing it into the bone matrix. **Osteoclasts** are mature bone cells that aid in remodeling and direct healing of bone. Osteoclasts remove calcium from bone when the levels of calcium in the body become too low. Bone lining cells lie on the outside of bone and aid in moving calcium and phosphates in and out of the bone matrix.

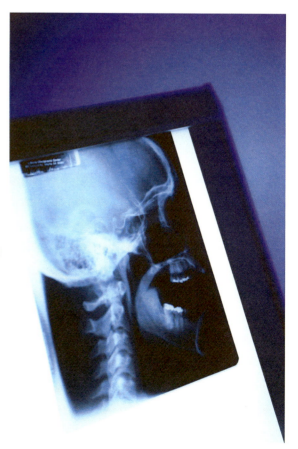

Compact Bone Organization

Bone tissue is a highly vascular tissue. Once the osteocyte is encased in the tissue it must receive its nutrients from blood. Blood enters the bone through the periosteal blood vessels. These vessels lie just below the periosteum. They then enter into the bone through perforating canals also called Volkmann's canals which traverse through the compact bone until it reaches the central canal. This blood must reach the central canal of each osteon which allows the blood to access the entire bone tissue. This central canal is also called the Haversian canal. From the central canal, which is the central portion of the osteon, the blood and nutrients flows out into the osseous tissue through small canals called canaliculi. This network of small canals allow for each osteocyte to receive its nutrients.

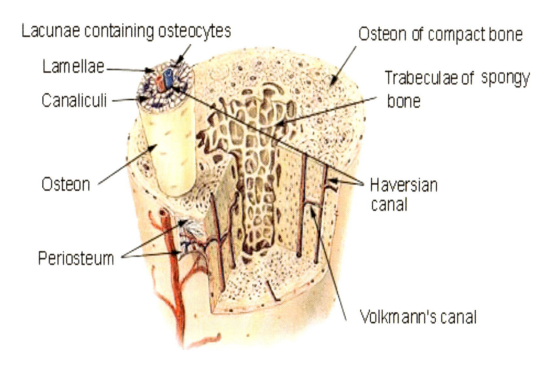

Compact bone arranged in concentric rings that radiate out from the **central canal (Haversian Canal)**. These rings are called **Lamellae**. The **canaliculi** must traverse through these rings to allow for the communication of each osteocyte that is housed in the **lacunae**.

The Osteon is the functional unit of compact bone.

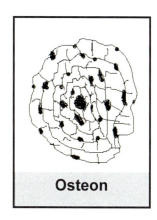

Bone Growth

Growth of bone occurs in a two step process. During fetal development most of our large bones are created in a process called **Intramembranous Ossification.** This process utilizes a cartilage membrane that will ossify into bone. Long bone growth called **Endochondral Ossification** is a type of bone growth beginning during the end stages of fetal development and occurs as active bone growth of the skeleton into adulthood. This growth begins with cartilage tissue that is calcified and ossified into bone tissue. Long bone growth begins as **Chondrocytes** (*cartilage cells*) in the center of the model start the process of hypertrophy as minerals are deposited into the matrix. This is known as the calcification phase of bone growth. Once calcified, the flow of nutrients to the chondrocytes fades away and the chondrocytes die. **Osteoblasts** move into the tissue and begin to secrete **osteoid**, an enzyme that hardens the organic component of bone. A periosteal bone collar forms around the cartilage of the bone model which is then surrounded by the periosteum. The periosteal bud, which consists of osteoblasts and blood vessels, invades the center of the cartilage model which will become the primary ossification center. This growth continues as new bone growth begins at the ends of the long bone creating secondary ossification centers creating spongy (trabeculated) bone. Once these secondary ossification centers are established the **Epiphyseal Plate** continues to create new bone on each end, lengthening the bone until maturity. This Epiphyseal Plate then closes creating the **Epiphyseal Line**, which is a remnant of what once was fully developing bone.

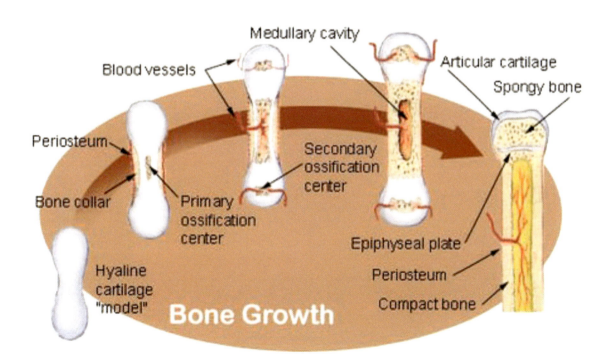

Bones of the Skeleton

Part of the Body	Name of the Bone	Part of the Body	Name of the Bone
Skull Cranium (8 bones) Face (14 bones) Ear (6 bones)	 Frontal (1) Parietal (2) Temporal (2) Occipital (1) Sphenoid (1) Ethmoid (1) Nasal (2) Maxillary (2) Zygomatic (2) Mandible (1) Lacrimal (2) Palatine (2) Inferior Conchae (2) Vomer (1) Malleus (2) Incus (2) Stapes (2)	**Upper Extremities** Pectoral Girdle Arm, Wrist, Hand	 Clavicle (2) Scapula (2) Humerus (2) Radius (2) Ulna (2) Carpals (16) Metacarpals (10) Phalanges (28)
Spinal column Vertebrae (26 bones) Hyoid Bone Sternum and Ribs	 Cervical Vertebrae (7) Thoracic Vertebrae (12) Lumbar Vertebrae (5) Sacrum (1) Coccyx (1) Sternum (1) True Ribs (14) False Ribs (10	**Lower Extremities** Pelvic Girdle Thigh, Leg, Ankle, Foot	 Os Coxae (2) Ilium, Ischium, and Pubis Femur (2) Patella (2) Tibia (2) Fibula (2) Tarsals (14) Metatarsals (10) Phalanges (28)

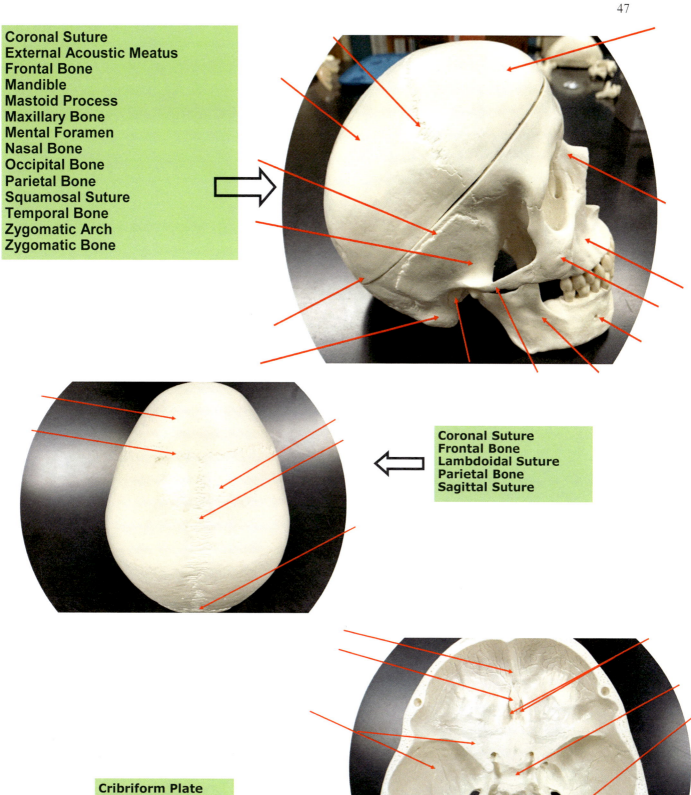

Ethmoid Bone
Frontal
Lacrimal Bone
Mandible
Maxilla
Nasal Bone
Nasal Turbinate (conchae)
Sphenoid Bone
Vomer
Zygomatic Bone

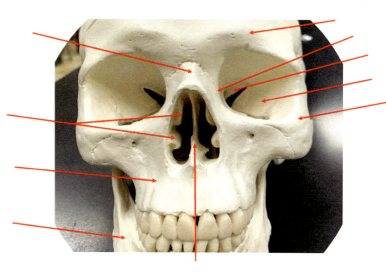

External Acoustic Meatus
Mandible
Mastoid Process
Styloid Process
Zygomatic Arch

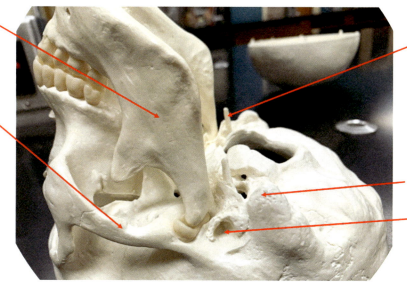

Foramen Magnum
Mandible
Mastoid process
Occipital Bone
Palatine Bone
Styloid Process
Zygomatic Arch
Vomer

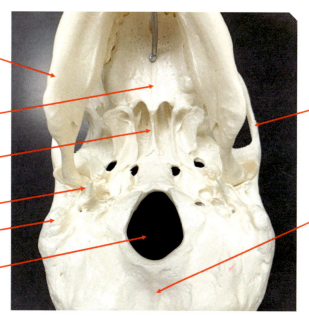

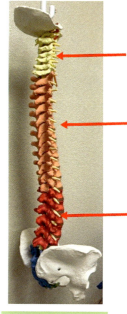

Cervical Vertebra
Lumbar Vertebra
Thoracic Vertebrae

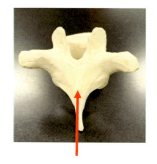 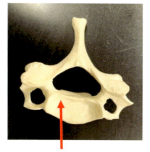 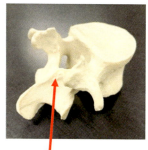

Atlas (C1)
Axis (C2)
Hyoid

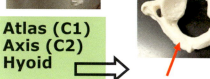

Number of Cervical Vertebrae	_____
Number of Thoracic Vertebrae	_____
Number of Lumbar Vertebrae	_____

Body of Vertebrae
Lamina
Pedicle
Spinous Process
Transverse Foramen
Transverse Process
Vertebral Foramen

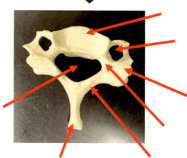

Body (Gladiolas)
Manubrium
Sternum
Xyphoid Process

Coccyx
Sacrum

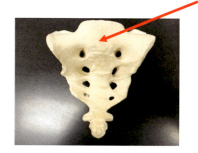 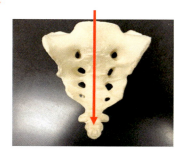

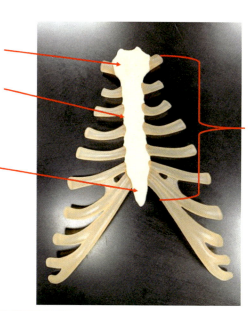

Number of True Ribs	_____
Number of False Ribs	_____
Number of Floating Ribs	_____

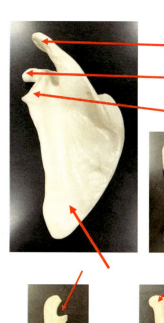

Pectoral Girdle:

Acromion Process
Clavicle
Coracoid Process
Glenoid Cavity (fossa)
Scapula

Humerus:
Capitulum
Greater Tubercle
Head of Humerus
Lesser Tubercle
Trochlea

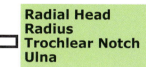

**Radial Head
Radius
Trochlear Notch
Ulna**

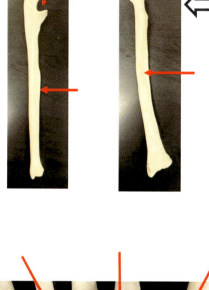

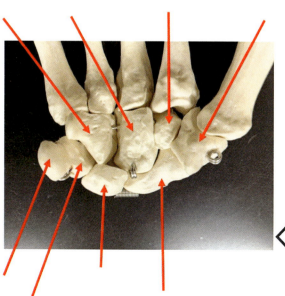

**Capitate
Hamate
Lunate
Pisiform
Scaphoid
Trapezium
Trapezoid
Triquetral**

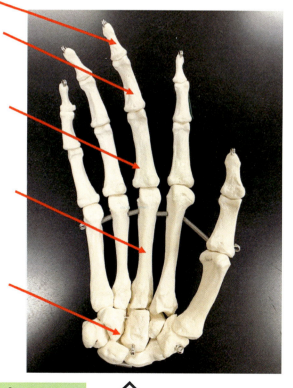

**Carpal Bones
Distal Phalanx
Metacarpals
Middle Phalanx
Proximal Phalanx**

Identify the Gender of these two Pelvises

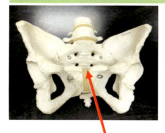
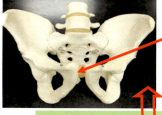

Pelvic Girdle:
Acetabulum
Female Pelvis
Iliac Crest
Ilium
Ischium
Ischial Tuberosity
Male Pelvis
Obturator Foramen
Pubis

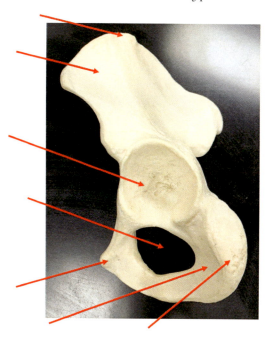

Lower Extremity:
Body of Femur
Femoral Head
Fibula
Greater Trochanter
Head of Fibula
Lateral Condyle
Lateral Malleolus
Lesser Trochanter
Medial Condyle
Medial Malleolus
Patella
Tibia
Tibial Tuberosity

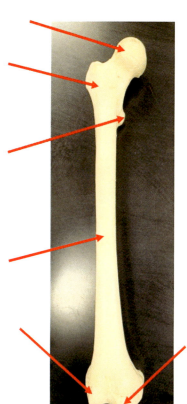

Tarsal Bones
- Calcaneus
- Cuboid
- Talus
- Medial Cuneiform
- Middle Cuneiform
- Lateral Cuneiform
- Navicular
- Metatarsals

Phalanges
- Distal Phalanx
- Middle Phalanx
- Proximal Phalanx

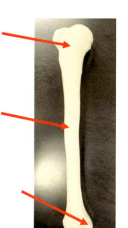

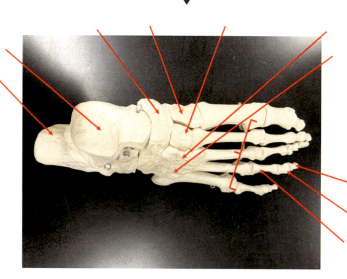

Muscle Tissue   Lab 8

The muscular system is designed for contraction. These contractions help the body to carry out the function of body movement. There are over 600 muscles that make up the human body and each is composed of muscle tissue, connective tissue, and nervous tissue. Skeletal muscle itself accounts for approximately 40% of the total weight of the body and its three main functions include movement, heat production, and posture/support.

Muscle tissue itself can be broken down into three main types. Smooth muscle, the type of muscle that is involuntary, found in the abdominal viscera. Smooth muscle has a stellate appearance and has its nucleus central to the cell. This muscle type is non-striated and its typical function is for peristalsis (*movement of food stuffs through the alimentary canal*). Cardiac muscle, just as the name implies is found in the heart. It looks very similar to skeletal muscle with two distinct differences. Cardiac muscle has a branching appearance and includes intercalated discs. Cardiac muscle is also involuntary. Skeletal muscle, the only voluntary muscle of the body, is primarily used for body movement. It appears as a non-branching striated tissue which can be multinucleated. The nuclei are pushed to the outer portion of the cell to allow for the contractile units of the muscle.

Smooth Muscle   **Cardiac Muscle**

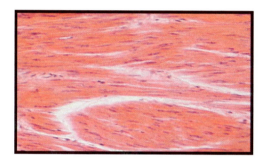 

Skeletal Muscle

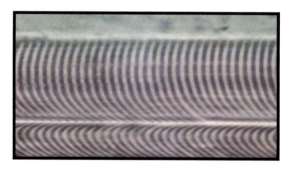

Although there are three types of distinct muscle, they all share common qualities. These qualities include **irritability** (*sensitive to neural stimuli*), **contractility** (*ability to shorten upon stimulation*), **extensibility** (*ability to extend when relaxed*) and **elasticity** (*an ability to stretch beyond their resting point*).

Skeletal muscle along with its binding connective tissue is arranged in a highly organized pattern. These patterns allow for contraction to allow movement, its main purpose.

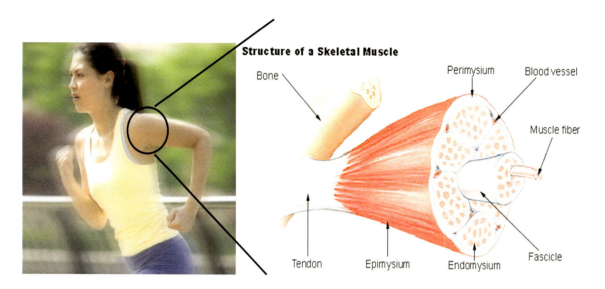

Skeletal Muscle

Skeletal muscles are attached to bones by tendons. There are two bony attachment sites, one is the **origin**, usually proximally located and less moveable than the **insertion** which is distal and always pulled toward the origin during muscle contraction. The fat fleshy portion of the muscle is called the muscle belly. Tendons can have different shapes and sizes. The flat sheet-like tendon found in the wrists and ankles is called the **aponeuroses**. Each tendon is wrapped in a protective covering called the **tendon sheath** and is bound to the boney attachment site which is on the outer periosteum of the bone itself. Muscle fibers join together to become bundles which are then grouped together into specific muscle fascicles. These fascicles are then grouped together to create specific muscles. This grouping of fibers and fascicles and eventually muscles is accomplished by structures created from connective tissue. The **endomysium** is a fine sheath, covering individual fibers which are bound together to form bundles called fasciculi covered by the **perimysium**. The **epimysium** covers the entire muscle and extends to become part of the tendon. The fibrous connective tissue that covers the musculature and attaches it to the skin is called **fascia**. The superficial fascia attaches skin to the underlying structures and can be thick or thin but is always laced with adipose tissue. Deep fascia an inward extension of the superficial fascia lacks adipose tissue and blends into the epimysium of the muscle. It is this deep fascia that combines adjacent muscles together, compartmentalizing and binding them into functional groups. Muscles often take different shapes which allow for a wide range of body activities. Muscles that run **parallel** are strap-like, long muscles that provide endurance for the body. **Convergent** musculature is fan shaped and when contracted, apply force into a single point. **Sphincteral** musculature have fibers that are arranged concentrically around an orifice and allow for constriction of the muscle around a fixed point, and **pennate** muscles are fan shaped and consist of many fibers and are short and strong.

Muscle Groups

Muscles don't usually work alone, just as fibers act within their bundles, muscles usually work with a specific group to perform an action. Muscles that contract together are called **synergists** and are usually closely related to each other. Muscles that oppose an action are called **antagonists** and are usually located opposite that of the muscles involved.

Muscle Fibers and Contraction

Skeletal muscle cells appear different from other cells of the body, but they are actually the same. They each contain Mitochondria, intracellular membranes, and organelles found in many of the cells of the body.

Skeletal muscle is organized in long cylindrical units called fibers. These fibers can range in size from as small as 10 µm in diameter running 6 cm in length to as large as 25 cm in length which can be seen easily with the naked eye. Skeletal muscle fibers can contain more than one nucleus and are as such considered multinucleate. These nuclei lie just below the outer portion of the cell itself called the **sarcolemma** (*plasma membrane of a muscle cell*). These nuclei are pushed to the outer portion due to the contractile elements of the cell. These contractile elements, **myofibrils** run parallel to one another and take up nearly the entire portion of the **sarcoplasm** (*cytoplasm of a muscle cell*). On each fibril are light and dark bands that alternate producing the striated appearance seen in skeletal muscle. **Myofibrils** are created from even smaller structures called **myofilaments**. These myofilaments are composed of two contractile proteins, **actin** (*thin filament*) and **myosin** (*thick filament*). These proteins slide together during contraction. The contractile unit of the muscle is called the sarcomere and extends from one **Z-line** to the next. Within the two Z-lines is the **A band** (*the dark band*), which contains both thick and thin filaments. Between each A Band lies the **I Band** (*the light band*) which contains only thin filaments. There are numerous sarcomeres within a muscle fiber. When a muscle contracts, it is the sarcomere which shortens in length, but the only part of the sarcomere that actually shortens is the I Band.

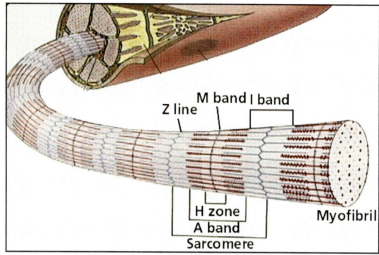

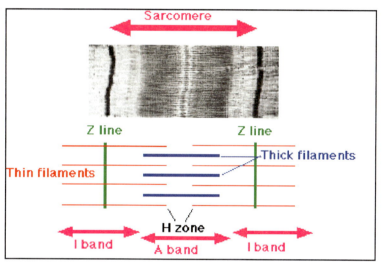

The filaments themselves do not shorten, but they actually slide over one another as contraction of the muscle fiber occurs. During this process the actin (*thin filaments*) on both sides of the A Band extend deeper into the center to increase the area where they intersect and slide over the myosin (*thick filaments*). At each junction of the A and the I bands, the sarcolemma folds into the cell forming a **transverse tubule** (*T Tubule*). These tubules run deep into the muscle cell between cross channels called **terminal cisternae**, which are part of the **sarcoplasmic reticulum** (*the smooth ER of the muscle cell*).

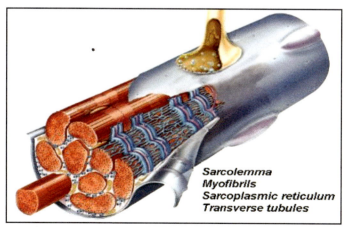

The Neuromuscular Junction

Afferent and **efferent** nerves innervate each muscle fiber. The efferent neuron is called the **motor neuron** and is responsible for contraction or relaxation of the muscle, and the afferent neuron is the **sensory neuron** which provides sensory input from the muscle cell. Each motor neuron has a threadlike axon extending from the CNS to a group of muscle fibers where it divides into numerous branches called **Axon Terminals (bulbs)**. These Axon Terminals attach to the sarcolemma of the muscle cell at the **Motor Endplate**, also called the **Neuromuscular Junction**. The neurotransmitter **Acetylcholine** is stored in the axon terminal which is released when the nerve is activated. It is this release of Acetylcholine that results in contraction of the muscle.

The Motor Unit

A **Motor Unit** consists of a single motor neuron and the muscle it innervates. All the fibers associated with the motor unit will contract due to the action of the motor neuron. In general there is about 1 motor unit for every 100-150 muscle fibers. In muscles that require greater dexterity, it may be 1 motor unit for every 10 fibers and for larger more massive muscles it can be 1 motor unit to every 500 muscle fibers.

Muscle Stimulation

Nerve impulses stimulate the muscle to contract. Once the motor neuron is stimulated it sends a signal to the motor endplate where synaptic vesicles release the neurotransmitter **Acetylcholine** into the synaptic cleft. Once enough Acetylcholine crosses the channel and is received by the receptors, sodium and potassium channels open to trigger depolarization of the **sarcolemma**. This signal is transmitted through the **T-Tubule** which stimulates the release of calcium stored in the **sarcoplasmic reticulum**. Calcium infuses into the muscle and binds with the protein **Troponin** which unlocks allowing the protein **Tropomyosin** to slide away from binding sites found on the **Actin** filament. With the actin binding site exposed the protein filament **Myosin** grabs the actin filament and pulls it toward the center creating contraction. This contraction continues until calcium is removed from the muscle returning the ion to the sarcoplasmic reticulum and shutting down the contraction.

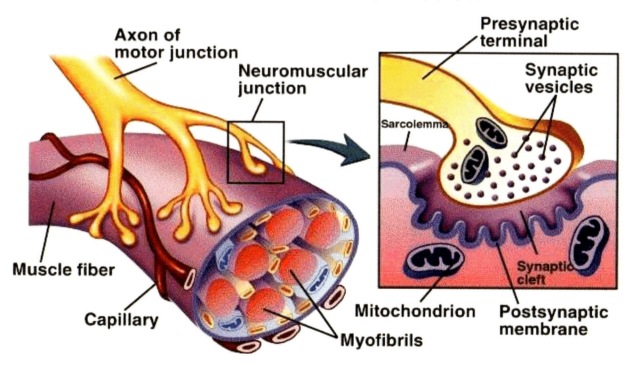

Review Exercise

Retrieve a slide of each of the three muscle types, smooth, cardiac, and skeletal muscle. Draw each slide identifying the differences between each type.

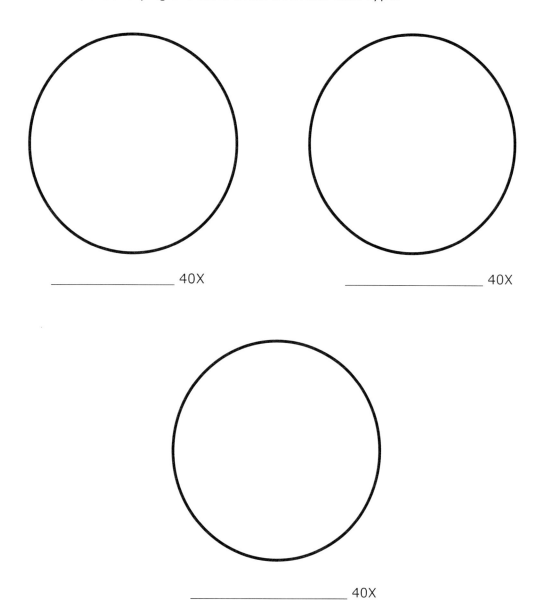

_____ 40X

_____ 40X

_____ 40X

Explain the distinguishing differences between each of the three muscle tissues.

Explain the differences between involuntary and voluntary musculature.

The diagram below represents the muscle myofibril. Using the terms below identify the structures indicated.

A) A Band   B) Actin (Thin filament)   C) I Band   D) Myosin (Thick Filament)
     E) Sarcomere   F) H Zone   G) Z Disc

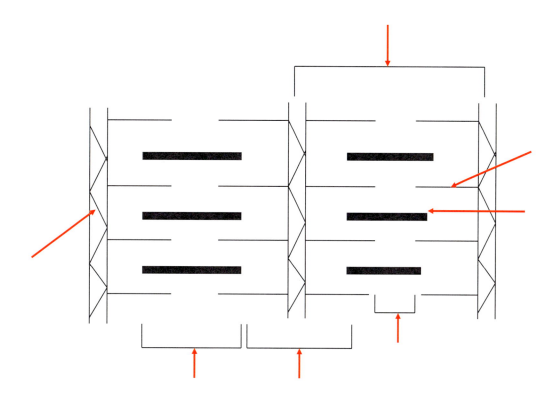

Explain the role of the Sarcomere:

Briefly outline muscle contraction:

Skeletal Muscle Identification   Lab 9

Identifying skeletal musculature is an enormous task to undertake. As the science of anatomy has progressed muscles have been named according to a number of differing criteria. **Direction** of muscle fibers is one of those criteria. Typically fibers that run parallel to the structure of the body it interfaces is called a **Rectus** muscle. These fibers or fascicles run straight along the length of that specific body structure. Those fibers that run across this structure are considered **Transverse** and those that run at an angle are considered **Oblique**. **Size** is another important aspect of nomenclature. Terms such as **Longus, Brevis, Minimis and Maximus** give us a good idea of muscle size. The **Number of Origination Sites** also can be included in the name, such as **Triceps** or **Biceps**. These muscles are named according to the number of heads that originate on specific bones. **Location** of the muscle on the body is another key determinant as seen with the **Tibialis Anterior**. This muscle is found on the anterior aspect of the Tibia. **Shape** can also play a key role in names. The **Trapezius** muscle is so named because it resembles a trapezoid. Lastly muscles are named according to their **Action** as in muscles that are **Extensors** or **Flexors** (**Extensor Digitorum** or **Flexor Pollicis Brevis**). Attempting to learn every muscle can seem like a huge undertaking, but following these easy naming rules the task may not be as difficult as you might think.

Muscles of the Head and Neck

When looking at the muscles of the head and neck we must first compartmentalize them into differing groups. Muscles of Facial Expression, Muscles of Mastication, or Muscles that move the eyes and those that move the head and neck.

Muscles of Facial Expression

These muscles have a major difference in that they do not insert onto a bone Instead these muscles insert into the superficial skin of the face allowing for a wide range of expression. Emotions are displayed on the face and is an important function in language.

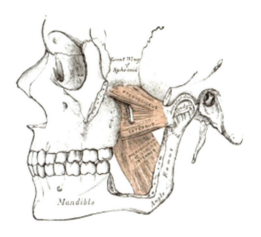

Muscles of Mastication

These muscles manipulate the food we eat in the chewing process, also called mastication. There are four major muscles of mastication. The Temporalis and the Masseter allow for the up and down motion of the mandible, while the Medial and Lateral Pterygoids allow for the side to side grinding motion of the jaw. These muscles work in tandem to break up food into smaller and smaller pieces for easy absorption within the alimentary canal.

Extrinsic Eye Muscles

There are six small muscles that move the eye. These muscles, Superior Rectus, Inferior Rectus, Medial Rectus, Lateral Rectus, Superior Oblique and the Inferior Oblique all move in coordination with the other to give a wide field of vision. When we look to the left our left Lateral rectus brings the left eye laterally, and yet the right eye is moved medially due to the Medial Rectus. Both eyes move together using the opposite musculature.

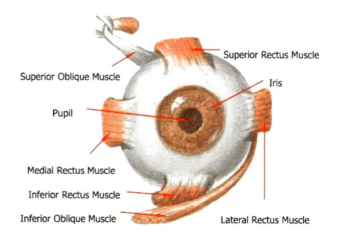

Muscles of the Head and Neck

Muscle	Origin	Insertion	Action

Epicranius
    Frontalis
 	
    Occipitalis
	Galea Aponeurotica
	
	Occipital, Temporal Bone
		Skin of Eyebrows, Root of Nose
		Galea Aponeurotica
			Raises Eyebrows
			
			Pulls Scalp Posteriorly

Orbicularis Oculi	Maxilla, Frontal	Circles the Eyes	Closes Eye
Orbicularis Oris	Maxilla, Mandible	Circles the Mouth	Closes Lips
Buccinator	Maxilla, Mandible	Angle of Mouth	Compresses Cheeks
Zygomaticus	Zygomatic Bone	Angle of Mouth Upper Lip	Elevates Angle of Mouth Upper lip
Levator Labii Superioris	Maxilla	Upper Lip, Nose	Elevates Upper Lip, Nose
Temporalis	Temporal Aspect of Skull	Mandible	Closes Jaw
Masseter	Zygomatic Arch	Mandible	Closes Jaw
Pterygoids	Inferior Aspect of Skull	Mandible	Medial—Closes Jaw Lateral—Opens Jaw
Sternocleidomastoid	Sternum, Clavicle	Mastoid Process	Rotates, Extends Head Secondary Muscle Respiration
Trapezius	Skull, Upper Vertebral Column	Scapula	Extends Head, Neck
Depressor Labii Inferioris	Mandible	Skin, Lower Lip	Draws Lip Inferiorly
Mentalis	Mandible	Skin of Chin	Protrudes Lower Lip, Wrinkles Chin

Identify the Marked Muscles

Buccinator
Depressor Labii Inferioris
Epicranius Frontalis
Levator Palpebrae
Masseter
Mentalis
Nasalis
Orbicularis Occuli
Orbicularis Oris
Scalenes
Sternocleidomastoid
Superior Corrugator
Temporalis
Trapezius
Zygomaticus Major/Minor

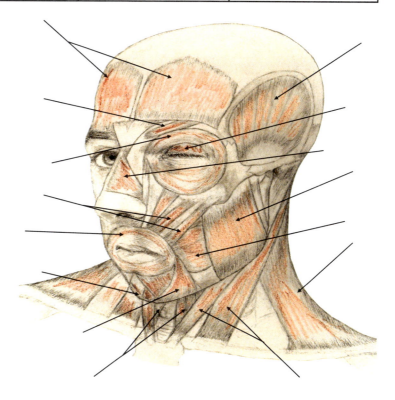

Muscle of the Trunk

Erector Spinae Group	Vertebrae, Pelvis	Superior Vertebrae, Ribs	Extends Back
Deep Back Muscles	Vertebrae	Vertebrae	Flexes or Extends Trunk
External Intercostals	Ribs	Superior Edge of Inferior Rib	Expands Thorax Primary Muscle Respiration
Internal Intercostals	Ribs	Inferior Edge of Superior Rib	Compresses Thorax Primary Muscle Respiration
Rectus Abdominis	pubis	Inferior Thoracic Cage	Flexes Waist
External Abdominal Oblique	Inferior Thoracic Cage	Midline of Abdomen	Compresses Abdomen
Internal Abdominal Oblique	Pelvis	Midline of Abdomen	Compresses Abdomen
Transversus Abdominis	Vertebrae, Pelvis, Ribs	Midline of Abdomen	Compresses Abdomen
Trapezius	Skull, Upper Vertebral Column	Scapula	Extends Head, Neck Rotates Scapula
Levator Scapulae	Vertebrae	Scapula	Elevates Scapula
Rhomboideus	Vertebrae	Scapula	Retracts Scapula
Serratus Anterior	Ribs (1-8)	Scapula	Protracts Scapula
Pectoralis Minor	Ribs (3-5)	Scapula	Depresses Scapula
Pectoralis Major	Ribs (1-6), Clavicle, Sternum	Humerus	Adducts, Flexes Arm
Teres Major	Scapula	Humerus	Extends, Adducts, Rotates Arm
Latissimus Dorsi	Vertebrae	Humerus	Extends Arm
Infraspinatus	Scapula	Humerus	Extends, Rotates Arm
Supraspinatus	Scapula	Humerus	Abducts Arm
Subscapularis	Scapula	Humerus	Extends, Rotates Arm
Teres Minor	Scapula	Humerus	Adducts, Rotates Arm
Deltoid	Scapula, Clavicle	Humerus	Abducts Arm
Diaphragm	Inferior Aspect of Rib and Sternum, Cartilage of Ribs, Lumbar Vertebrae	Central Tendon	Primary Inspiratory Muscle

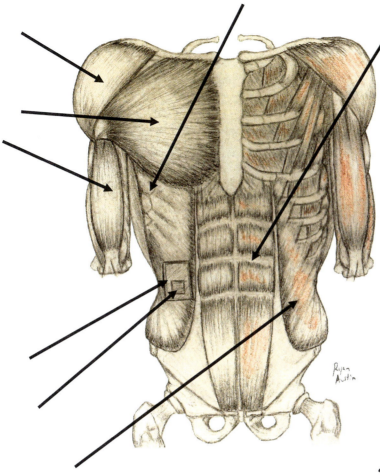

**Biceps Brachii
Deltoid
External Abdominal Oblique
Internal Abdominal Oblique
Pectoralis Major
Rectus Abdominus
Serratus Anterior
Transverse Abdominus**

**External Abdominal Oblique
Infraspinatus
Internal Abdominal Oblique
Latissimus Dorsi
Rhomboid Major
Semispinalis Capitis
Splenius Capitis
Sternocleidomastoid
Teres Major
Teres Minor
Trapezius**

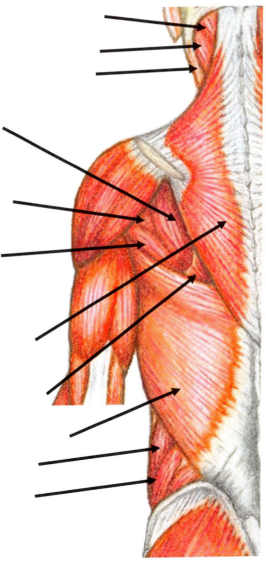

Muscles of the Upper Extremity

Muscle	Origin	Insertion	Action
Triceps Brachii	Humerus, Scapula	Ulna	Extends Forearm
Biceps Brachii	Humerus, Scapula	Radius	Flexes, Supinates Forearm
Brachialis	Humerus	Ulna	Flexes Forearm
Pronator	Ulna, Humerus	Radius	Pronates Forearm
Flexor Carpi Ulnaris	Medial Epicondyle of Humerus	Carpal Bone	Fleses, Abducts Wrist
Flexor Carpi Radialis	Medial Epicondyle of Humerus	Metacarpal Bone	Flexes, Abducts Wrist
Flexor Digitorum	Medial Epicondyle of Humerus, Ulna, Radius	Phalanges	Flexes Fingers
Brachioradialis	Humerus	Distal Radius	Flexes, Pronates Forearm
Supinator	Ulna	Radius	Supinates Forearm
Extensor Carpi Ulnaris	Lateral Epicondyle of Humerus	Metacarpal Bone	Extends, Abducts Wrist
Extensor Carpi Radialis	Lateral Epicondyle of Humerus	Metacarpal Bone	Extends, Abducts Wrist
Extensor Digitorum	Lateral Epicondyle of Humerus	Phalanges	Extends Fingers

Brachialis
Brachioradialis
Biceps Brachii
Deltoid
Flexor Carpi Radialis
Flexor Carpi Ulnaris
Flexor Digitorum
Pectoralis Major
Pronator Teres
Palmaris Longus
Triceps Brachii

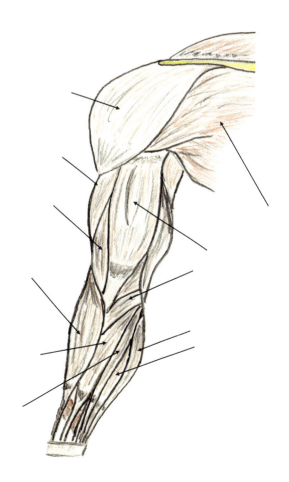

Muscles of the Lower Extremity			
Muscle	Origin	Insertion	Action
Iliopsoas	Ilium, Vertebrae	Femur	Flexes Thigh
Tensor Fasciae Latae	Hip	Tibia	Abducts Thigh

Gluteal Group
   Gluteus Maximus
   Gluteus Medius
   Gluteus Minimus
	
	Hip
	Hip
	Ilium
		
		Femur
		Femur
		Femur
			
			Extends Thigh
			Extends Thigh
			Abducts Thigh

Quadriceps Femoris
   Rectus Femoris
   Vastus Lateralis
   Vastus Medialis
   Vastus Intermedialis
	
	Ilium
	Femur
	Femur
	Femur
		
		Tibia
		Tibia
		Tibia
		Tibia
			
			Extends Leg, Flexes Thigh
			Extends Leg
			Extends Leg
			Extends Leg

Sartorius	Ilium	Tibia	Flexes Thigh, Flexes and Rotates Leg

Hamstring Group
   Biceps Femoris
   Semimembranosus
   Semitendinosus
	
	Ischium, Femur
	Ischium
	Ischium
		
		Fibula
		Tibia
		Tibia
			
			Flexes Leg, Extends Thigh
			Flexes Leg, Extends Thigh
			Flexes Leg, Extends Thigh

Adductor Group
   Adductor Longus
   Gracilis
	
	Pubis
	Pubis
		
		Metatarsal Bones
		Tibia
			
			Dorsiflexion of Foot
			Adducts Thigh

Tibialis Anterior	Tibia	Metatarsal Bones	Dorsiflexion of Foot
Extensor Digitorum Longus	Tibia	Phalanges	Extends Toes

Triceps Surae
   Gastrocnemius
   Soleus
	
	Femur
	Tibia, Fibula
		
		Calcaneus
		Calcaneus
			
			Plantar Flexes Foot
			Plantar Flexes Foot

Fibularis (Peroneus) Longus	Tibia, Fibula	Tarsal, Metatarsal Bones	Flexes, Everts Foot

Identify the marked structures of the anterior and posterior leg.

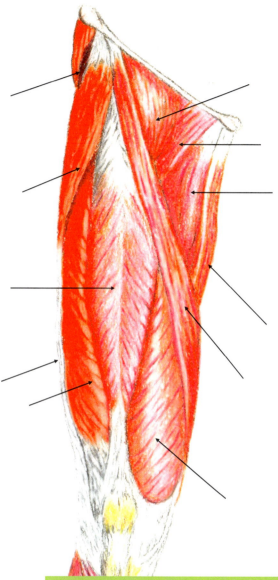
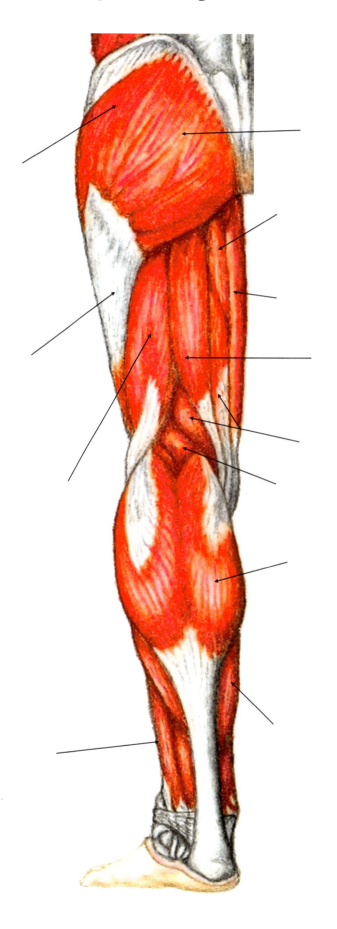

Identify the muscles of the lower extremity
Some of the Muscles can be found on both pictures
A) Adductor Longus
B) Adductor Magnus
C) Biceps Femoris
D) Gastrocnemius
E) Gluteus Maximus
F) Gluteus Medius
G) Gracilis
H) Iliopsoas
I) Illiotibial Tract
J) Pectineus
K) Peroneus (Fibularis) Longus
L) Popliteus
M) Rectus Femoris
N) Sartorius
O) Semimembranosus
P) Semitendinosus
Q) Soleus
R) Tensor Fasciae Latae
S) Vastus Lateralis
T) Vastus Medialis

Review Exercise

Based on the naming systems of muscles, identify the criteria used for naming the muscles listed below. (note: more than one criteria can be used)

_____Rectus Abdominis   a) Action of Muscle

_____Rhomboid Major   b) Location of Muscle associated with Landmarks

_____Gluteus Minimis   c) Size of muscle

_____Triceps Brachii   d) Shape of Muscle

_____Tibialis Anterior   e) Direction of Muscle Fiber

_____Flexor Digitorum   f) Number of Origination Sites

Briefly explain the differences between the origin and insertion sites of musculature.

Identify the muscles of mastication.

Identify the muscles of respiration, (*include primary and secondary musculature*).

Identify the muscles of the Triceps Surae.

Which muscles compose the Quadriceps Group?

Which muscles compose the Hamstrings Group?

Nervous tissue   Lab 10

The nervous system is composed of two major divisions: the **central nervous system** (CNS) and the **peripheral nervous system** (PNS). The central nervous system encompasses the brain and the spinal cord. The peripheral nervous system include all the nerves that transmit impulses to and from the central nervous system. Often the nervous system is divided into afferent and efferent divisions. The **afferent** division includes the nerves that send signals to the brain, while the **efferent** division send signals from the brain to the tissues of the body. The afferent division includes all of the sensory nerves and sensory tracts. The efferent division include all the motor nerves and motor tracts. The efferent division can also be divided into the **somatic nervous system** (*innervates the skeletal system*) and the **autonomic nervous system** (*innervates the visceral organs*).

Cells of the Nervous System

The neuron is the basic cell of nervous tissue. It conducts impulses to and from the CNS. **Neuroglia cells** are special cells that nourish and support the neurons. **Neuroglia cells outnumber neurons 10 to 1.** The neuroglia cells include astrocytes, oligodendrocytes, microglial cells, Schwann cells, satellite cells, and ependymal cells.
Astrocytes cover the blood vessels and they combine to help create the **blood brain barrier.** Substances that pass in and out of the blood vessel must cross this barrier formed by the astrocytes. **Oligodendrocytes** form a layer around the axons in the CNS creating a **myelin sheath**. This sheath eliminates signal loss increasing the speed in which impulses

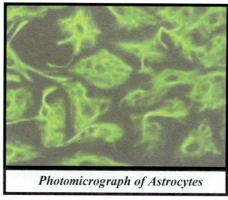
Photomicrograph of Astrocytes

move to and from the neuron. **Microglial cells** help damaged or infected tissue through phagocytic activity. **Satellite Cells** are protective cells and surround nuclei found outside the central nervous system while the **Schwann** cells (*also called neurolemocytes*) form the myelin sheaths around axons in the PNS. **Ependymal cells** line the fluid filled spaces of the brain. They contain cilia which help to circulate the cerebrospinal fluid they create.

Neuron

The neuron itself is composed of three main structures. The cell body, also called the soma is the enlarged area of the cell which contains the cytoplasm, nucleus and organelles called **Nissl Bodies**. The body forms a cone shaped structure called the **axon hillock** where the axon projects out of the body. The axon itself is wrapped in the myelin sheaths created by the appropriate glial cell and conducts impulses away from the cell body. The myelin sheaths in the PNS, created from schwann cells are in segments. Each segment is separated by a **Node of Ranvier**. Impulses appear to jump from one node to the next through a process called **saltatory conduction**. Extending out from the cell body are branching fibers called **dendrites**. These branched extensions are sensitive to other neurons. The dendrites form a synapse or connection between two or more neurons, and the axon conducts the impulse along. Most of the sensory neurons are unipolar in structure.

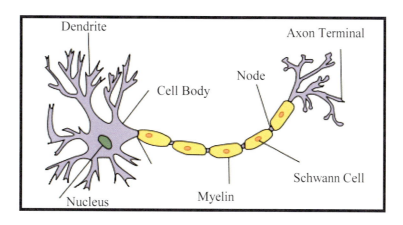

Neurons can be classified as multipolar, bipolar, and unipolar. Multipolar neurons have multiple projections and compose most of the neurons in the CNS. Most of the motor neurons are multipolar neurons. Bipolar neurons have two projections from the cell body. These neurons are commonly found in the sensory tracts for sight and smell. Unipolar neurons have one extension that divides into two distinct branches, one branch receives stimuli through dendritic endings and the other acts as an axon conducting impulses along. Most of the unipolar neurons are sensory neurons as well.

Nerve Structure

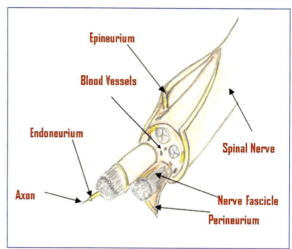

Each nerve fiber is enclosed in a connective tissue sheath called the **endoneurium**. This wrapping insulates the individual fiber from other fibers running adjacent to it. Groups of fibers are bound together into a nerve fascicle and surrounded by a thicker sheathing called the **perineurium**. The entire nerve is then wrapped and bound together along with blood and lymphatic vessels by a white fibrous tissue sheath called the **epineurium**. Like neurons, nerves are classified according to the direction in which a signal impulse travels. If a nerve carries both sensory (*afferent*) and motor (*efferent*) fibers it is considered a mixed nerve. All of the spinal nerves are mixed nerves. Those nerves that carry only sensory fibers are referred to as sensory (*afferent*) nerves while those carrying only motor fibers are considered motor (*efferent*) nerves.

Neuronal Communication

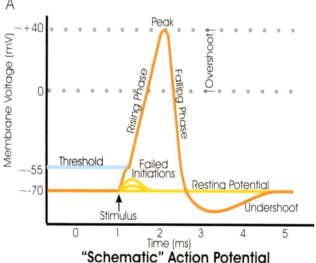

"Schematic" Action Potential

Neurons have two physiological properties, **excitability**, the ability to receive and respond to stimuli and converting them into nerve impulses, and **conductivity**, the ability of the neuron to send the signal to other neurons. In a resting neuron the outer surface of the membrane is slightly more positively charged than the inner surface. This difference in charge on the two sides of the membrane results in a voltage charge across the membrane. This is called the **resting membrane potential**. At this time the neuron is considered polarized. In this resting state the intracellular ion is predominantly potassium (K^+). The Sodium (Na^+) ion is the predominant ion of the extracellular fluid. When a neuron receives a signal that is strong enough to excite it into firing, the neuron becomes slightly more permeable to sodium (Na^+). These sodium ions rush into the cell increasing the positive ion concentration inside the cell and reversing its polarity. The inner surface of the neuron is now more positive than the outer surface of the cell and it is considered **depolarized**. When the change in polarity is strong enough it initiates an **action potential**. Within milliseconds after the influx of sodium, the membrane permeability changes. As a result the sodium permeability decreases and a permeability to potassium (K^+) increases allowing potassium to rush out of the cell and restore the original membrane potential. And is referred to as **repolarization**. Once an action potential is established it is a self-propagating phenomenon that spreads rapidly over the entire length of the neuron and is an all or nothing response to stimuli. This propagation of signal is also called the nerve impulse.

Neural Communication

Communication between neurons occurs at the synaptic cleft. The axonal bulb of the presynaptic neurons contain tiny vesicles filled with neurotransmitters. Each axon bulb is separated from the postsynaptic neuron by a tiny gap called the synaptic cleft. When an impulse reaches the axon terminal some of the vesicles rupture, releasing the neurotransmitter into the cleft. The neurotransmitter diffuses across the synaptic cleft to bind to membrane receptors on the next neuron initiating the axon potential.

Fused Synaptic Vesicle
Mitochondria
Neurotransmitter
Neurotransmitter receptor
Postsynaptic neuron
Presynaptic Neuron
Synaptic Cleft
Synaptic Vesicle

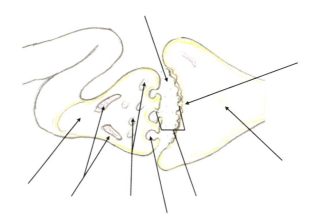

Microscope Exercise

Retrieve a neuron slide from the slide box and draw the neuron in the space provided

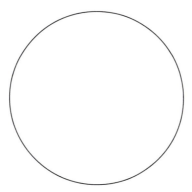

Review Exercises

Briefly describe the differences between the intracellular fluid and the extracellular fluid around the neuron while it is at rest.

Review Exercises

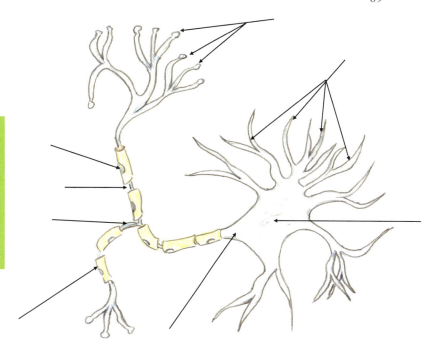

Identify the marked structures

Axon Bulb
Axon Hillock
Axon of Neuron
Dendrite
Myelin Sheath
Node of Ranvier
Nucleus of Cell Body
Schwann Cell (Neurolemocytes)

Briefly explain what is meant by the statement that an action potential is an all or nothing response.

Briefly explain Saltatory Conduction

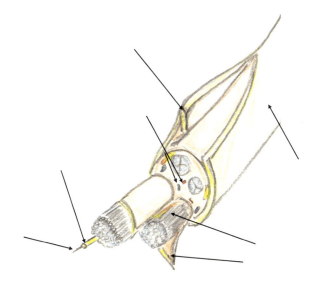

Identify the marked structures

Axon
Blood Vessels
Endoneurium
Epineurium
Nerve Fascicle
Perineurium
Spinal Nerve

The Brain   Lab 11

The brain is housed in the cranial cavity. Between the bone and the brain are the **meninges**. These tissues help to cushion the brain and protect it from harm. The spinal cord is housed in the vertebral bodies and is bathed in **cerebrospinal fluid**. This fluid circulates from the ventricles of the brain into the spinal column. The brain, the main component of the central nervous system is composed of gray and white matter. The gray matter is composed of nerve cell bodies and dendrites, while the white matter is composed of myelinated tracts and axons. An adult brain weighs approximately 3 to 3.5 lbs and accounts for approximately 2% of body weight. Although it is a small organ it is highly specialized and requires a great deal of blood supply. Somewhere between 15-25% of all cardiac output to the body is sent to the brain.

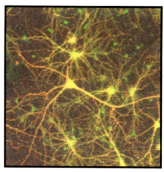

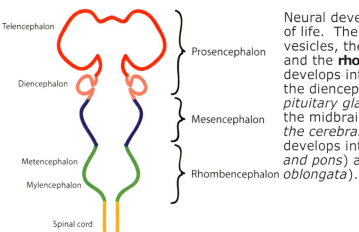

Neural development begins on about the 17th day of life. The brain itself is derived from three primary vesicles, the **prosencephalon**, the **mesencephalon**, and the **rhombencephalon**. The Prosencephalon develops into the telencephalon (*the cerebrum*) and the diencephalon (*the thalamus, hypothalamus, and pituitary gland*). The mesencephalon develops into the midbrain (*the superior, inferior colliculi and the cerebral peduncles*) and the Rhombencephalon develops into the metencephalon (*the cerebellum and pons*) and the Myelencephalon (*the medulla oblongata*).

The cerebrum is the largest and most obvious portion of the brain. It accounts for approximately 80% of brain mass. It is responsible for higher mental functions and reasoning. It has two hemispheres which are separated by a longitudinal fissure. They are connected by a mass of axonal fibers called the **Corpus Callosum**, a large tract of white matter. Each hemisphere contains a lateral ventricle that is filled with cerebrospinal fluid and each controls different aspects of a persons mental abilities, including analytical and verbal skills (*reading/writing/speaking*), spatial and artistic skills (*creative thought*). The corpus callosum unifies the two hemispheres allowing for shared learning and memory. The cerebrum itself consists of two layers, a surface layer called the cerebral cortex,

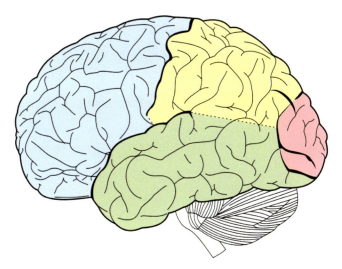

(*composed mostly of gray matter*) and a deeper layer composed mostly of white matter. The cerebral cortex is characterized by numerous folds and grooves in the tissue called convolutions. These form from early fetal development and are thought to allow more tissue to be placed in a smaller space. They actually allow for a tripling of the surface area for gray matter. The elevated folds of the convolutions are called **cerebral gyri**, while the depressed grooves are called **cerebral sulci**.

Lobes of the Cerebrum

The **frontal lobe** forms the anterior portion of each cerebral hemisphere. A prominent deep furrow called the **central sulcus** separates the frontal and the parietal lobe. This central sulcus runs at a right angle from the longitudinal fissure to the lateral sulcus which separates the frontal and the temporal lobes. The **precental gyrus** is immediately in front of the central sulcus and is responsible for the motor portions of the brain. The frontal lobe is responsible for initiating voluntary motor impulses, analyzing sensory experiences and providing responses relating to personality. It is also related to memory, emotions, reasoning, judgment, planning and verbal communication.

The **parietal lobe** lies posterior to the frontal lobe. The anterior portion contains the **postcentral gyrus**, which is the main sensory area of the brain. It is designated as a somatesthetic area because it responds to stimuli from cutaneous and muscular receptors of the body. As with the precentral gyrus, the postcentral gyrus does not correspond in size to the part of the body being served. A good understanding of this can come from looking at the **homunculus,** a detailed look at the amount of brain that serves different areas of the body. The precentral gyrus deals with the motor areas and the postcentral gyrus is associated with the sensory. The parietal lobe also helps with understanding speech and articulating thoughts and emotions. It also interprets textures and shapes of objects being handled.

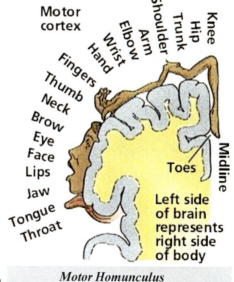

Motor Homunculus

The **temporal lobe** is located below the parietal lobe and posterior to the frontal lobe. It contains the auditory centers that receive sensory fibers from the cochlea of the ear. This lobe also interprets some sensory experiences and stores memories of both auditory and visual experiences.

The **occipital lobe** forms the posterior portion for the cerebrum. It is not distinctly separated from the parietal and temporal lobes and lies superior to the cerebellum. It is separated from the cerebellum by the **tentorium cerebelli** and functions with visual input. It integrates eye movement by directing and focusing the eye. This lobe is also responsible for visual association, correlating visual images with previous visual experiences and other sensory stimuli

The **insula** is a deep lobe of the cerebrum that cannot be seen on the surface. It lies deep to the lateral sulcus and is covered by portions of the parietal and temporal lobes. There is not much known about the insula, but it has been shown to help integrate other cerebral activities. It is thought to help with memory as well.

White Matter

The deeper portion of the cerebrum consists of dendrites, myelinated axons, and associated neuroglia. It is here that billions of connections are made. There are three types of fibers or tracts within the white matter and are named according to the location and the direction in which they conduct impulses. **Association fibers** are confined to a given cerebral hemisphere, they conduct impulses between the neurons within that specific hemisphere. **Commissural fibers** connect the neurons and gyri of one hemisphere to those of the other. The **corpus callosum**, and the **anterior commissure** are composed of commissural fibers. **Projection fibers** form the ascending and descending tracts that transmit impulses from the cerebrum to other parts of the brain and spinal cord.

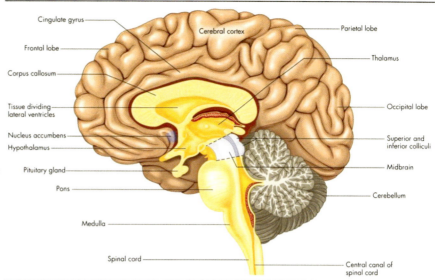

Speech Centers

The speech area is located in the left inferior gyrus of the frontal lobe. Wernicke's area, located in the superior gyrus of the temporal lobe is directly connected to Broca's through the arcuate fasciculus. Wernicke's aphasia is when one speaks real words, but the words are jumbled or mixed and spoken randomly. If Wernicke's area is destroyed one cannot understand either written or spoken language. If Broca's area is damaged one cannot speak at all. The knowledge of language in the brain has been learned through the study of these aphasias.

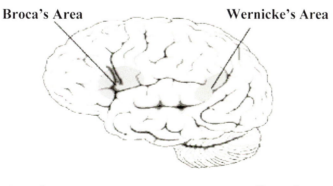

Basal Nuclei of the Cerebrum

The basal nuclei are paired masses of gray matter located deep within the white matter of the cerebrum. The most prominent is the corpus striatum. It contains several nuclei within its structure. The **caudate nucleus**, the **lentiform nucleus** and the **claustrum**. These nuclei are associated with other structures particularly in the mesencephalon. The caudate and putamen control unconscious contractions of skeletal muscle. The Globus Pallidus regulates muscle tone and is necessary for specific intentional body movement.

Diencephalon

The **diencephalon** is the major autonomic region of the brain and consists of the thalamus, hypothalamus, epithalamus and the pituitary gland. This is the second subdivision of the telencephalon. The **thalamus** is a paired organ positioned below the lateral ventricle. Its principle function is to act as a relay center for all sensory impulses, except smell, to the cerebral cortex. The thalamus plays an important role in the initial autonomic response to pain and therefore is partially responsible for the physiological shock that follows trauma. The hypothalamus named for its position below the thalamus is the most inferior portion of the diencephalon. It acts on the autonomic nervous center. It either is an accelerator or decelerator. It secretes several hormones which are then funneled to the pituitary gland through the infundibulum. It also regulates cardiovascular activity, temperature, electrolyte balance, hunger, sleeping and wakefulness, sexual response, emotions and endocrine function. The epithalamus is located in the posterior portion of the diencephalon. It forms the roof of the third ventricle. The inside lining of the roof consists of the vascular choroid plexus which creates cerebrospinal fluid. The epithalamus connects the left and right superior colliculi. The pituitary gland is a small rounded pea shaped gland and is also called the cerebral hypophysis. It is protected from damage by the boney structure, the Sella Turcica located on the superior aspect of the Sphenoid bone. This gland is the major endocrine gland of the body and is divided into the anterior (*adenohypophysis*) and a posterior (*neurohypophysis*) lobes.

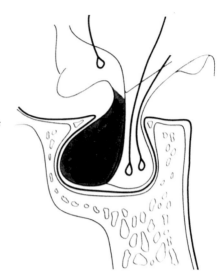

Mesencephalon

The **mesencehalon** contains the **corpora quadrigemina** a structure concerned with visual and auditory reflexes. This area also contains the cerebral peduncles, composed of fiber tracts, as well as specialized nuclei for maintaining posture and movement. The midbrain contains the mesencephalic aqueduct (*aqueduct of sylvius*) which connects the third and fourth ventricles. The corpora quadrigemina are four rounded elevations of the posterior portion of the midbrain. The upper two are the **superior colliculi** concerned with visual reflexes, and the inferior two, the **inferior colliculi** are responsible for auditory reflexes. The cerebral peduncles are a pair of cylindrical structures composed of ascending and descending projection fiber tracts that support and connect the cerebrum to the other regions of the brain. The red nucleus lies deep in the midbrain and functions in reflexes of motor coordination and maintenance of posture. The substantia niagra inhibits forced involuntary movements. Its color reflects the high content of melanin pigment.

Metencephalon

The **metencephalon** contains the **pons** and the **cerebellum**. The pons consists of white fibers that course in two directions. The surface fibers run transversely to connect the cerebellum through the middle cerebellar peduncles. The deeper fibers run longitudinally and are part of the motor and sensory tracts that connect the medulla oblongata with the tracts of the midbrain. Within the pons are several nuclei, cranial nerve V (*Trigeminal*), cranial nerve VI (*Abducens*), cranial nerve VII (*Facial*) as well as cranial nerve VIII the (V*estibulocochlear*). There are other nuclei located here that aid in breathing.

The Cerebellum

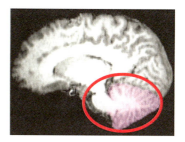

The cerebellum, the second largest structure of the brain, consists of two hemispheres and a central constricted area called the **vermis**. The falx cerebelli is the portion of the meninges that partially extends between the hemispheres. Like the cerebrum it has a superficial layer called the cerebellar cortex and a thick deeper layer composed of white matter. The deeper white mater has a distinctive branching pattern called the arbor vitae (*due to its tree like appearance*) which can be seen on the sagital view. The principle function of the cerebellum is coordinating skeletal muscle contractions. It is aided by the proprioceptors in the muscles, tendons and joints which are sensitive to changes in tension.

Medulla Oblongata

The medulla oblongata connects directly to the spinal cord and houses many of the cranial nerves and vital autonomic centers. It is the most inferior portion of the brainstem. It houses the fourth ventricle which is continuous with the central canal of the spinal cord.

Meninges of the Central Nervous System

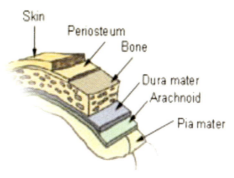

Meninges of the CNS

The three main meninges of the nervous system are the **Dura Mater**, the **Arachnoid** and the **Pia Mater**. The Dura Mater is in direct contact with the bone of the skull. It adheres the brain to the skull. The Dural sinuses are located in the Dura and they collect venous blood and drain it into the internal jugular veins of the neck. The Dura forms the dural sheath that extends into the vertebral canal and surrounds the spinal cord. The Arachnoid is the middle layer of the three meninges. It is a delicate netlike membrane that spreads over the CNS but does not extend into the sulci or fissures of the brain. The subarachnoid space is located between the arachnoid mater and the deeper pia mater and contains cerebrospinal fluid. The Pia Mater is a thick layer which is tightly bound to the convolutions of the brain and the irregular contours of the spinal cord. The Pia mater helps to nourish the brain and spinal cord and forms part of the choroid plexus in the ventricles of the brain.

Ventricles of the Brain

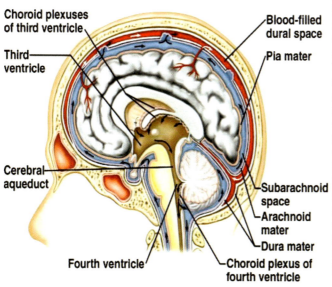

The ventricles, central canal and subarachnoid space contain cerebrospinal fluid which is formed by the active transport of substances from blood plasma in the choroid plexus. The ventricles are connected to one another along with the central canal of the spinal cord. The lateral ventricles are located in each corresponding hemisphere below the corpus callosum. The interventricular foramen which connects the lateral ventricles to the third ventricle is located in the diencephalon. The mesencephalic aqueduct connects the third ventricle to the fourth ventricle. CSF exits the fourth ventricle into the subarachnoid space through three foramina, a median aperture and two lateral apertures. The brain weighs approximately 1500 grams, but when suspended in CSF its weight decreases to about 50 grams. Anywhere from 500-800 ml of CSF is created each day, composed of proteins, glucose, urea, and white blood cells.

Spinal Cord

The spinal cord is a continuation of the brain and actually ends near the first Lumbar Vertebra. There are two prominent enlargements, a cervical enlargement and a lumbar enlargement. These areas service the upper and lower extremities. The inferior portion of the spinal cord contains many nerves that extend beyond the cord itself. This structure is called the cauda equine. It is named appropriately due to its appearance (*a horses tail*). The spinal cord develops as 31 segments. Each segment gives off a pair of spinal nerves. When the cord is cut transversely you can see white matter surrounding a central gray H-like structure. This structure has two anterior and two posterior horns. The posterior aspect of the spinal cord receives the sensory tracts while the anterior aspect provides the motor tracts.

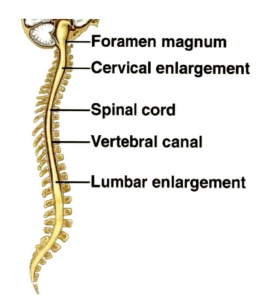

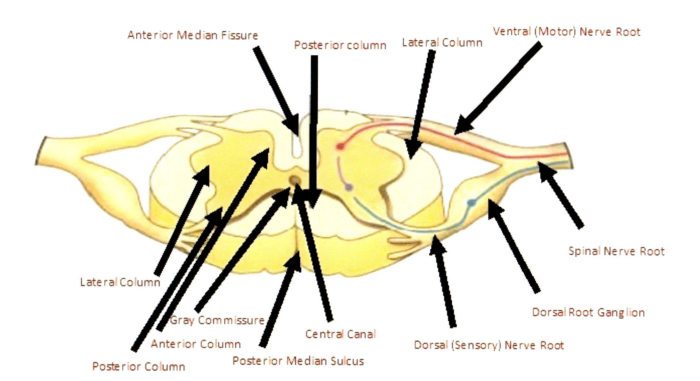

Brain Dissection

1) Collect a preserved sheep brain from your instructor. If the brain is encased in the meninges, carefully cut through the outer tough covering and remove the brain while trying to preserve the cranial nerves and pituitary gland. Examine the brain for the following structures:

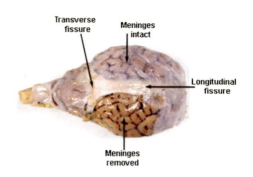

Frontal Lobe **Olfactory Bulb**
Temporal Lobe **Optic Chiasma**
Parietal Lobe **Longitudinal Fissure**
Occipital Lobe **Sulci**
Cerebellum **Gyri**
Pons **Brainstem**

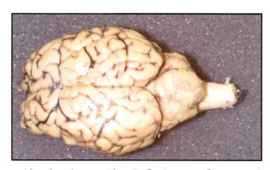 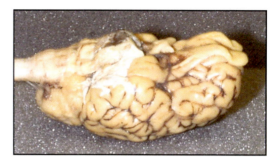

2) Place the brain on the inferior surface and slowly pull the cerebellum away from the cerebrum. As you gently pull it away inspect just below the cerebrum for the Inferior and Superior Colliculi. These combine to form the Corpora Quadrigemina.

3) Make a median sagital cut through the longitudinal fissure and cut the brain into left and right hemispheres.

4) Examine the interior aspect for the following structures:

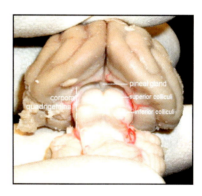

Arbor Vitae
Cerebelum
Cerebrum
Cerebral Aqueduct
Corpus Callosum
Fornix
Hypothalamus
Inferior Colliculi
Lateral Ventricle
Optic Chiasma
Pineal Gland
Pons
Septum Pellucidum
Superior Colliculi
Thalamus

4) Place the two hemispheres back together and make a coronal (*frontal*) cut dividing the brain into anterior and posterior halves. Examine the following structures:

Cerebral Cortex   White Matter

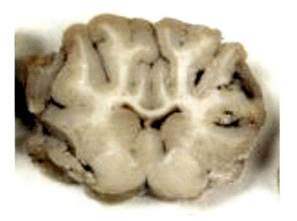

Review Exercise

Identify the following marked structures on the sheep brain.

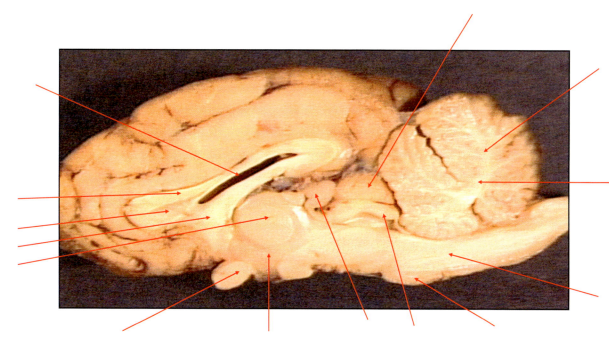

Arbor Vitae
Brain Stem
Cerebellum
Corpus Callosum
Fornix
Hypothalamus
Inferior Colliculi
Lateral Ventricle

Optic Chiasma
Pineal gland
Pons
Septum Pellucidum
Superior Colliculi
Thalamus

Review Exercise

Briefly discuss the difference between gray matter and white matter.

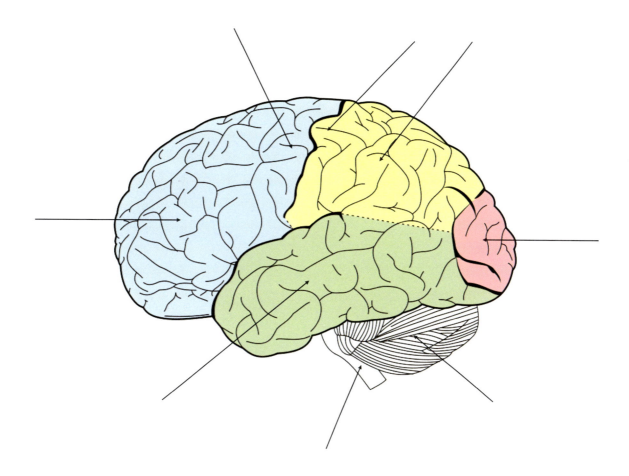

Identify the marked structures of the brain

Brainstem
Cerebellum
Frontal Lobe
Occipital Lobe
Parietal Lobe
Precentral Gyrus
Postcentral Gyrus
Temporal Lobe

Cranial Nerves

Lab 12

The portion of the nervous system outside of the Central Nervous System helps to convey impulses away from and to the brain and the spinal cord. These nerves are either Cranial Nerves or spinal nerves depending upon where they arise. Those that arise directly from the brain are considered Cranial nerves. These nerves are always referred by their number which ranges from 1 to 12, and always are listed as Roman Numerals. These twelve nerves are located in order from the anterior aspect of the brain to the posterior aspect. Most of these nerves are mixed nerves, but some are expressly sensory. It has now been discovered that there are actually only 11 true cranial nerves, Cranial Nerve 11 has been proven to be an actual derivative of Cranial Nerve 10.

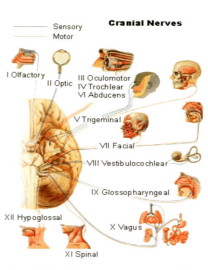

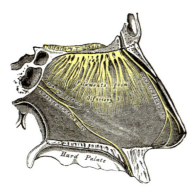

Cranial Nerve I

The Olfactory nerve consists of bipolar neurons that function as chemoreceptors. The dendrites and cell bodies are positioned within the mucosa which primarily cover the nasal conchae and are adjacent to the septum. The neurons pass through the Cribiform plate that lie on either side of the Christa Gali of the Ethmoid bone to the olfactory bulb where they are conveyed to the olfactory center of the cortex called the Uncus. These nerves do not synapse through the thalamus.

Cranial Nerve II

The optic nerve conveys sensory information to the brain from photoreceptors in the retina of the eye. Each optic nerve has an estimated 1.25 million nerve fibers that converge at the back of the eyeball and enter the cranial cavity through the otpic canal. Each eye has its own nerve that travels back to meet at the optic chiasma. Nerve fivers that arise from the medial portion of each eye cross (*decussate*) to the opposite side of the brain while the lateral portion runs straight back and continue on their perspective side. The optic nerves then pass to the thalamus via the optic tracts into thalamic nuclei where they converge in the occipital lobe as the visual cortex.

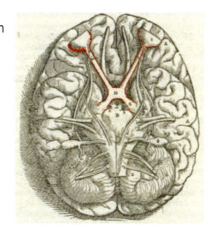

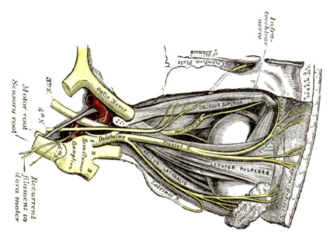

Cranial Nerve III

The Oculomotor nerve is primarily a motor nerve which produces extrinsic and intrinsic movement of the eyeball. It divides into superior and inferior nerve branches as it passes through the superior orbital fissure. The superior branch innervates the superior rectus and the levator palpebrae muscles, while the inferior branch innervates the medial and inferior rectus, as well as the Inferior oblique muscle. This nerve also has parasympathetic innervation to the eye for controlling lens shape and pupil constriction.

Cranial Nerve IV

The **Trochlear Nerve** is a small mixed nerve that arises from the midbrain and passes through the superior orbital fissure to innervate the superior oblique musculature which moves the eye down and toward the midline.

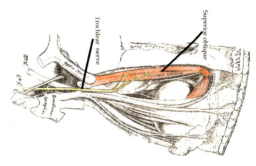

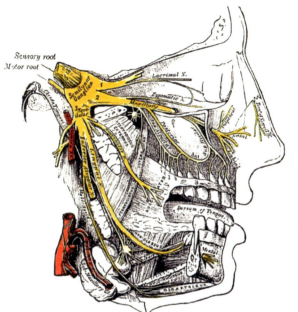

Cranial Nerve V

The Trigeminal nerve is a large mixed nerve the emerges from the anterolateral side of the pons. A large sensory root enlarges and is called the trigeminal ganglion which gives rise to three divisions, the ophthalmic, the maxillary and mandibular nerves. The smaller root contains motor fibers that accompany the mandibular nerve to innervate the muscles of mastication including the medial and lateral pterygoids, the masseter and the temporalis. The three sensory nerves respond to touch, temperature and pain from the face.

Cranial Nerve VI

The Abducens is a small nerve that originates from a nucleus within the pons and emerges from the its lower portion along the anterior border of the medulla oblongata. It is a mixed nerve and innervates the lateral rectus to move the eyes laterally.

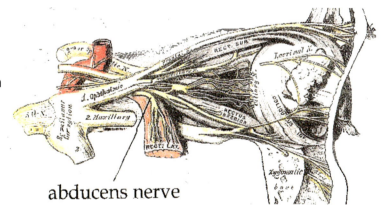

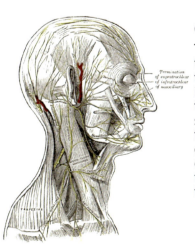

Cranial Nerve VII

The Facial nerve arises from nuclei within the lower portion of the pons. It is a mixed nerve which has motor pathways to the digastric muscle as well as the muscles of facial expression including the scalp and Platysma. It has autonomic functions to the submandibular, sublingual and the lacrimal gland. Sensory fibers arise from taste buds on the anterior two-thirds of the tongue which are chemoreceptors. The Geniculate ganglion is the enlargement of the facial nerve just before the entrance of the sensory portion in the pons. Sensations of taste are conveyed within the medulla to the thalamus and on to the gustatory center in the parietal lobe.

Cranial Nerve VIII

The Vestibulocochlear nerve is also know as the auditory or acoustic nerve. It is the only cranial nerve that does not leave the cranium through a foramen, but through a meatus. It is a combination of two parts, a vestibular nerve associated with 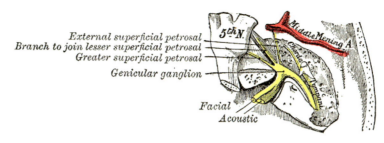 equilibrium and balance and a cochlear nerve associated with hearing. Fibers run from the inner ear through the internal acoustic meatus to enter the pons.

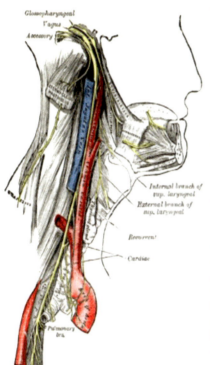

Cranial Nerve IX

The Glossopharyngeal nerve is a mixed nerve. It innervates part of the tongue and the pharynx. The motor nerve innervate the pharynx and the parotid gland to stimulate the swallowing reflex and to secrete saliva. The sensory fibers arise from the pharyngeal region, the parotid gland, the middle-ear cavity and the taste buds of the posterior one third of the tongue. Some sensory fibers also arise from the sensory receptors in the carotid sinus and help to regulate blood pressure.

Cranial Nerve X

The Vagus is the longest of all the cranial nerves. It is a mixed nerve and innervates the visceral organs of the thoracic and abdominal cavities. The motor portion arises from the nucleus ambiguous and the dorsal motor nucleus of X which is in the medulla and passes through the jugular foramen. It innervates the pharynx, larynx, respiratory tract, lungs, heart and the esophagus. It innervates most of the abdominal viscera except the lower portion of the large intestine. A branch off of the vagus nerve, the recurrent laryngeal nerve, enables speech.

Cranial Nerve XI

In Classical training the Accessory nerve was considered a motor nerve innervating the musculature for movement of the head, neck an shoulders. Initially conceived as a unique nerve, it was thought to arise from both the brain and the spinal cord. It has now been proven to be a derivative of Cranial Nerve X, the Vagus nerve. The cranial root arises from nuclei in the medulla and passes through the jugular foramen to innervate muscles of the soft palate, pharynx and the larynx for swallowing. The spinal root arises from the first five segments of the cervical spinal column and passes through the foramen magnum to join its cranial root and then through the jugular foramen to innervate the sternocleidomastoid and Trapezius musculature.

Cranial Nerve XII

The Hypoglossal nerve is a mixed nerve that arises from the hypoglossal nucleus within the medulla to innervate the extrinsic and intrinsic muscles of the tongue. Motor function allow for movements for food manipulation, swallowing and speech. Sensory aspects function for proprioception of the tongue.

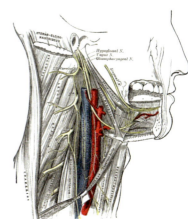

Lab Exercise
Using the worksheet below, test each of the cranial nerves on a lab partner.

Cranial Nerve Testing

Cranial Nerve I _____
The sense of smell. Test partner with each of the fragrances

Cranial Nerve II _____
The sense of sight. Test partner with Snellen eye chart on wall. Also check the pupil responses.

Cranial Nerve III_____
Cranial Nerve IV _____
Cranial Nerve VI _____
Eye movement. Test partner using the, H in Space.
Muscles that move the eye _____
_____
_____
_____
_____
SO 4, LR 6, All Else 3

Cranial Nerve V _____
General Sensation to the Face. Pin prick to each of the divisions
Divisions _____
_____
_____

Cranial Nerve VII _____
Muscles that move the face (Facial Expression)
Taste to _____

Cranial Nerve VIII _____
Sense of Hearing and Balance. Test Partner with tuning forks and Balance tests

Cranial Nerve IX _____
Ability to swallow
Taste to _____

Cranial Nerve X _____
Major nerve to the Visceral Organs
Listen for Borborygmi (Bowel Sounds)

Cranial Nerve XI _____
Ability to move head, neck and shoulders. Test partner with elevation And depression of shoulders as well as strength of the SCM Muscles

Cranial Nerve XII_____
Movement and strength of tongue.
Tongue movement ☐straight   ☐to left   ☐to right

Identify the marked Cranial Nerves/Nerve Structures

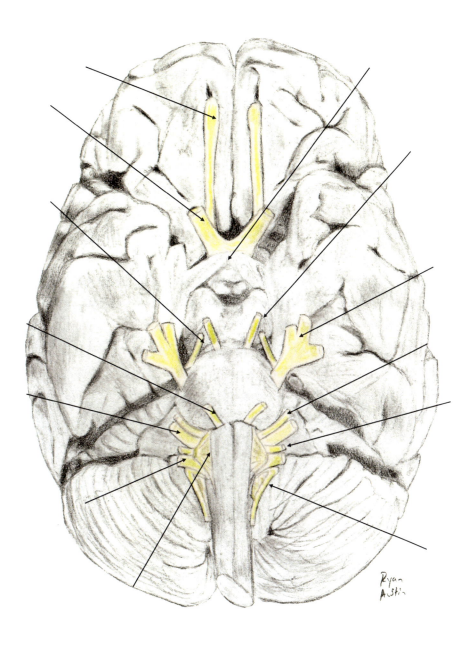

Abducens
Accessory
Facial
Glossopharyngeal
Hypoglossal
Oculomotor
Olfactory

Optic Nerve
Optic Chiasm
Trigeminal
Trochlear
Vagus
Vestibulocochlear

Sensory Organs   Lab 13

The sensory organs are highly specialized extensions of the nervous system. They contain sensory neurons adapted to respond to specific stimuli and conduct nerve impulses to the brain. Whenever a sensory impulse is transmitted to the brain it brings about an awareness of the stimulus, sometimes referred to as a sensation. The interpretation of this sensation is referred to as a perception. These sensations include impulses from tactile stimuli, auditory stimuli, visual stimuli, gustatory and olfactory stimuli. Our senses are classified as either general or special according to the degree of complexity of the receptors and their neural pathways. They are also classified as either somatic or visceral depending on the location of their receptors. General senses are widespread throughout the body and are structurally simple. These include sensations of touch, pressure, temperature and pain. Special senses are localized in complex receptor organs and have extensive neural pathways in the brain. These include taste, smell, sight, hearing and balance. Somatic senses are those located in the body wall and include the cutaneous receptors of the muscles, tendons and joints. The visceral receptors are found in the visceral organs of the body. Sensory neurons can also be classified according to their location and proximity to the environment around us. Exteroreceptors are those located near the surface of the body where they respond to stimuli from the external environment. Photoreceptors, mechanoreceptors, chemoreceptors, tactile receptors, thermoreceptors and nociceptors are of this classification. Visceroreceptors are those that produce sensations from the viscera. These can include internal pain, hunger, thirst, fatigue and nausea. Specialized receptors in the blood stream are sensitive to blood pressures and are called Baroreceptors. Information about body position, equilibrium and movement are produced by the Proprioceptors. These receptors are located in the inner ear, around joints and between tendons and muscles.

Tactile and Pressure Receptors

These receptors are sensitive to mechanical forces that distort or displace the tissue in which they are found. Tactile receptors respond to light touch and are located primarily in the dermis and the hypodermis of the skin. Pressure receptors respond to changes in pressure, vibration and stretch and are found in the hypodermis of the skin and in the tendons and ligaments of joints. Meissner's Corpuscles are mechanoreceptors and detect light motion against the skin surface. Free nerve endings can be tactile receptors, thermoreceptors and pain receptors. They detect touch, pressure, temperature and tissue damage. Pacinian Corpuscles are also mechanoreceptors and detect deep pressure and high frequency vibration. Another mechanoreceptor, the organ of Ruffini, detects deep pressure and stretch. And the Bulbs of Krause detect light pressure and low frequency vibration.

Two Point Tactile Discrimination Exercise

Retrieve two Sharpies of differing colors. Using your lab partner as a subject, ask them to close their eyes. The experimenter touches the palm of the subjects hand with a Black sharpie. With the subjects eyes still closed, ask the subject to touch the exact same location on his had with a red sharpie. Measure the difference in millimeters. This difference is the are of localization. Repeat the experiment twice more and average the results.

Area of Localization _____

Proprioceptors
These receptors advise the brain of our own movements by responding to changes in stretch and tension of the joints and musculature of the body. Proprioception information is used to adjust the strength and timing of muscle contractions to produce coordinated movement. Our Kinesthetic sense allows us to know where are body parts are without visual cues, such as dressing or walking in the dark. The four types of Proprioceptors include the joint kinesthetic receptor, the neuromuscular spindle, the neurotendinous receptor and the sensory hair cell

Olfaction
Olfactory receptors are the dendritic endings of the olfactory nerve that respond to chemical stimuli and transmit the sensation of smell directly to the olfactory portion of the cerebral cortex. This sense is not highly developed in humans due to our ability to depend on our other senses. It is considered the least important sense of the body. The olfactory receptors are located in the nasal epithelium of the roof of the nasal cavity on both sides of the nasal septum. The free ends have several dendritic endings called olfactory hairs that respond to airborne molecules that enter the nasal cavity. The sense of olfaction has several neural segments. Unmyelinated axons of the olfactory cells unite to form the olfactory bulbs that lie on both sides of the crista galli on the ethmoid bone. Within the olfactory bulbs the nerves synapse with dendrites of the olfactory tract which then transmit the impulses to the olfactory center in the cerebral cortex..

Gustation
The sense of taste comes from specialized epithelial receptor cells. These cells are Clustered together as taste buds. They respond to chemical stimuli and transmit our taste sense through the glossopharyngeal, facial, and Vagus nerves. As noted in the cranial nerve section of the nervous system, the anterior 2/3rds of the tongue sensory input is channeled through the facial nerve while the posterior 1/3rd travels over the glossopharyngeal nerve. But is worth mentioning that the Vagus nerve plays a very minor role in taste. The impulse ultimately enters the taste center located in the parietal lobe of the cerebral cortex. Each cell contains a dendritic hair cell that protrudes through the bud in a taste pore. Saliva provides the moistened environment necessary for chemical stimuli to activate the gustatory cell. Taste buds are elevated by surrounding connective and epithelial tissue which form papillae. There are four papillae, Vallate, Fungiform, Foliate, and Filiform. Vallate papillae are the largest but least numerous and are arranged in an inverted V-shape pattern on the back of the tongue. Fungiform are knoblike papillae that are present on the tip and sides of the tongue. Foliate are distributed on the sides of the tongue and are the most sensitive. And the Filiform are short, thick and threadlike. They are present on the anterior two-thirds of the tongue. Taste buds are only found in the Vallate, Fungiform, and Foliate papillae. The Filiform papillae, while the most numerous lack a taste hair cell. The four basic tastes are sweet, sour, bitter, and salty. The umami receptor was once thought to sense metallic tastes, but now is considered an enhancer for all tastes. Currently taste centers found for the four basic tastes are a highly debatable subject, but some believe that the sweet center is found on the tip, sour on the sides, bitter toward the back, and salty all over the tongue.

Taste Exercise
Identify the four basic tastes and map the location of each on the picture provided.

Basic Tastes _____

_____

_____

_____

Vision

Rods and cones are the photoreceptors within the eye that are sensitive to light energy. They are stimulated to transmit nerve impulses through the optic nerve and optic tract to the visual cortex of the occipital lobe. The eyes are organs that refract and focus incoming light rays onto sensitive photoreceptors at the back of each eye.

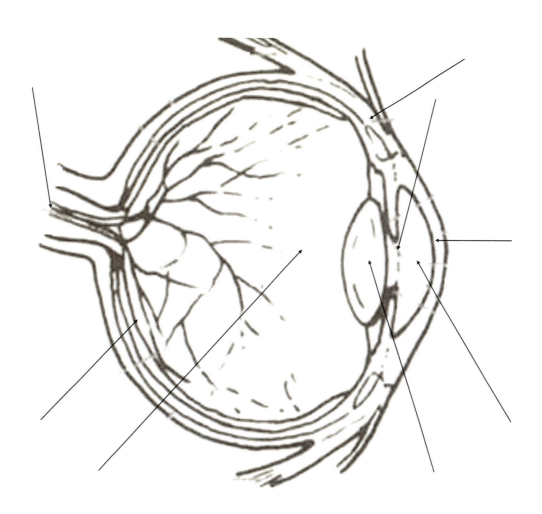

Identify the marked structures

Cornea
Lens
Sclera
Retina
Optic Nerve
Anterior Chamber
Posterior Chamber
Pupil

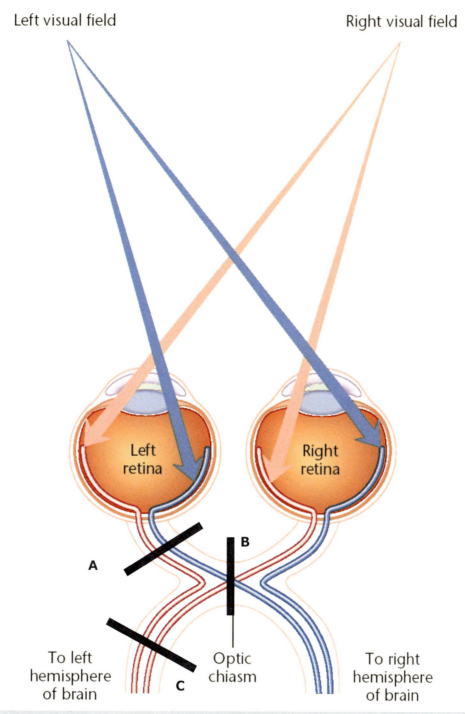

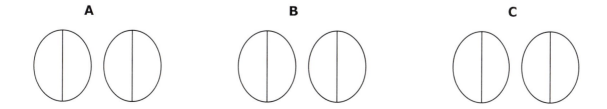

Cow Eye Dissection

1) Obtain a preserved cow eye from your instructor.

2) Examine the external surface of the eye, noting the thick cushion of adipose and muscle tissue. Gently cut the excess tissue from around the eye. Identify the optic nerve that exits the eye on the posterior aspect.

3) Holding the eye with the cornea facing downward, carefully make an incision with a sharp scalpel into the sclera about 6mm above the cornea. (note: the sclera is very tough and when you may need to use substantial pressure to make the incision.) Once completed use the scissors to cut around the eye removing the cornea from the posterior aspect.

4) Carefully lift the anterior part of the eye away from the posterior. You should see the vitreous humor remaining in the posterior aspect.

5) Examine the anterior aspect and identify the following:
   Ciliary Body-black pigmented body that appears to be a halo circling the lens

   Lens-Biconvex structure that is opaque in preserved specimens

6) Carefully remove the lens and identify the adjacent structures.
   Iris-Anterior continuation of the Ciliary body penetrated by the pupil

   Cornea-More convex anterior portion of the sclera (normally transparent But cloudy in preserved specimens)

7) Examine the posterior portion of the eye. Carefully remove the vitreous humor and identify the following.
   Retina-the neural layer of the retina appears as a delicate yellowish-white, probably crumpled membrane that separates easily from the pigmented choroids.

   Note its point of attachment. What is this point called?_____

*Note: the pigmented Choroid coat appears iridescent due to a special reflecting surface called the **tapetum lucidum**. This specialized surface reflects light within the eye is found in the eye of animals that live under condition of low-intensity light. This iridescent color is not found in humans.*

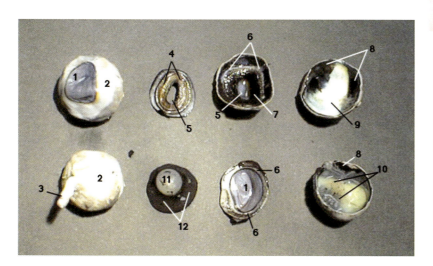

_____ Choroid Coat
_____ Ciliary Muscle
_____ Cornea
_____ Iris
_____ Lens
_____ Optic Nerve
_____ Pupil
_____ Retina
_____ Sclera
_____ Suspensory Ligaments
_____ Tapetum Lucidum
_____ Vitreous Humor

Writing a Lab Report

In this course your will be required to report your findings in a written format. This lab report should encompass all of the information needed as well as the information gained from completing the lab assignment. The report should include all the elements of a scientific research paper. It should have a cover page which includes the title of the experiment, the author's name, course name and instructor. It should consist of five sections: Introduction; Materials and Methods; Results; Discussions and Conclusions; and References.

Sample Report Layout

Lab Report

Cover Page
    Title of experiment
    Author's Name
    Course
    Lecture Instructor
    Lab Instructor

Introduction (*Everything you know about the subject matter*)
    Provide detailed background information
    Explain everything in easy to understand language
    State your Hypothesis clearly and concisely

Material and Methods
    List all supplies used in the experiment
    Give a detailed explanation of how you conducted the experiment

Results
    Report data (can be in any fashion, graphs, tables and drawings)
    Clearly summarize your findings

Discussion and Conclusions
    Clearly analyze your data
    Conclude if data gathered supported your hypothesis
    Explain any uncontrolled variables or difficulties

References
    Cite any sources used for your report

Cover & Contents: Vitruvian Man: Leonardo Da Vinci: Public Domain
1. Ghost Image: Muscle Man: Microsoft Clip Art
2. Anatomical Planes: GNU Free Documentation License
3. Body Regions: NCI Original: Public Domain
4. Small Image: Muscle Man: Microsoft Clip Art, Large Image: Original
6. Large Image: GNU Free Documentation License, Skull Images: Microsoft Clip Art
7. Small Image: Original
8. Body Cavities: NCI Original: Public Domain
9. Atom: Microsoft Clip Art
10. pH Scale: Original
11. Vial: Microsoft Clip Art
12. Periodic Table: Original
14. 15. 16. Microscope Pictures: Microsoft Clip Art
17. Both Images: Microsoft Clip Art
18. Mitochondria: *http://micro.magnet.fsu.edu,DNA: Clipart*
19. Slide Use: Microsoft Clip Art
20. Cell Division: en.wikipedia: Licensing GDFL
     Mitosis Cycles: Original NIH: Public Domain
22. Cell: Original
23. Mitosis: Originals, Plasma Membrane: http://micro.magnet.fsu.edu
35. Compact Bone: Original Cardiac Muscle: Original
36. Hair Pic: Istockpoto
38. Finger: Original
39. Skin Layers: en.wikipedia: Licensing GDFL
41. All Bone Pictures: en.wikipedia: Licensing GDFL
42. X-Ray: Microsoft Clip Art
44. Compact Bone: en.wikipedia: Licensing GDFL, Osteon: Original
45. Bone Growth en.wikipedia: Licensing GDFL
47-51. Skeleton and Bone Pictures : Original
52. Muscle Slides: en.wikipedia: Licensing GDFL
53. Adapted from Public Domain Images
54. Muscle Structures : en.wikipedia: Licensing GDFL
55. Neuromuscular Junction: www.classontheweb.com
57. Muscle Filament: Original, Muscle structure: en.wikipedia: Licensing GDFL
58. Jaw: Grey's Amatomy Public Domain, Eye: en.wikipedia: Licensing GDFL
59-64. Original
66. Astrocyte: www.ocu.Mit.edu, Neuron en.wikipedia: Licensing GDFL
67. Nerve/Action Potential: en.wikipedia: Licensing GDFL, Original
68. Synapse: en.wikipedia: Licensing GDFL
69. Neuron: Original, Nerve: en.wikipedia: Licensing GDFL
70-75.. Brain Picture: en.wikipedia: Licensing GDFL
76-77. Original
78. Brain Image: Public Domain
79-81. Cranial Nerve Pictures: Grays Anatomy Public Domain
83. Brain: Public domain
84. Senses: Microsoft Clip Art
86. Tongue and Eye en.wikipedia: Licensing GDFL
87. Visual Fields: en.wikipedia: Licensing GDFL
88. Eye Disection: Original

All narrative work is original, based upon 18 years of teaching Anatomy and Physiology. © 2021